W.J. Loudon

An elementary treatise on rigid dynamics

W.J. Loudon

An elementary treatise on rigid dynamics

ISBN/EAN: 9783742892478

Manufactured in Europe, USA, Canada, Australia, Japa

Cover: Foto ©berggeist007 / pixelio.de

Manufactured and distributed by brebook publishing software
(www.brebook.com)

W.J. Loudon

An elementary treatise on rigid dynamics

AN ELEMENTARY TREATISE

ON

RIGID DYNAMICS

BY

W. J. LOUDON, B.A.

DEMONSTRATOR IN PHYSICS IN THE UNIVERSITY OF TORONTO

" $\mu\eta\delta\grave{\epsilon}\nu$ $\mathring{a}\gamma a\nu$ "

New York

THE MACMILLAN COMPANY

LONDON: MACMILLAN & CO., Ltd.

1896

PREFACE.

THIS elementary treatise on Rigid Dynamics has arisen out of a course of lectures delivered by me, during the past few years, to advanced classes in the University.

It is intended as a text-book for those who, having already mastered the elements of the Calculus and acquired some familiarity with the methods of Particle Dynamics, wish to become acquainted with the principles underlying the equations of motion of a solid body.

Although indebted to the exhaustive works of Routh and Price for many suggestions and problems, I believe that the arrangement of the work, method of treatment, and more particularly the illustrations, are entirely new and original; and that they will not only aid beginners in appreciating fundamental truths, but will also point out to them the road along which they must travel in order to become intimate with those higher complex motions of a material system which have their culminating point in the region of Physical Astronomy.

My thanks are due Mr. J. C. Glashan of Ottawa, who has kindly read the proofsheets and supplied me with a large collection of miscellaneous problems.

W. J. LOUDON.

UNIVERSITY OF TORONTO, Aug. 19, 1895.

CONTENTS.

CHAPTER V.

CHAPTER VI.

CHAPTER VII.

CHAPTER VIII.

CHAPTER IX.

CHAPTER X.

.

RIGID DYNAMICS.

———∘∘⦂⊗⦂∘∘———

CHAPTER I.

MOMENTS OF INERTIA.

1. In attempting to solve the equations of motion of a *Rigid Body* in a manner similar to that employed for a single particle, it will be found that certain new quantities appear, which depend on the extent and shape of the body, on its density, and on the way in which it may be moving in respect of some particular line or system of coördinate axes.

2. These quantities are called *Moments of Inertia* and *Products of Inertia*. A moment of inertia of a body about any line is defined to be the sum of the products of all the material elements of the body by the squares of their perpendicular distances from this line. It may be denoted in general by the letter I, and when I is expressed in the form MK^2, where M is the mass of the body, K is called the *radius of gyration*. When the body is referred to three coördinate rectangular axes, the moments of inertia about the three axes will evidently be

$$A = \Sigma m(y^2 + z^2), \quad B = \Sigma m(z^2 + x^2), \quad C = \Sigma m(x^2 + y^2),$$

m being the mass of any element at the point (x, y, z), and the summation being taken throughout the body.

A product of inertia is defined with reference to two planes at right angles to one another and is found by multiplying the elements by the products of their distances from these coördi-

B   I

nate planes, and summing them throughout the body.   Products
of inertia exist in sets of three, and for three rectangular axes
are
$$D = \Sigma myz, \quad E = \Sigma mzx, \quad F = \Sigma mxy.$$

3. It is evident that when the law of m is known and the
shape of the body is given, the finding of a moment or of a
product of inertia involves an integration ; and the following
examples will serve to show how the process of integration may
be used for this purpose.   Further on, several propositions will
be given by which the method may be usually much simplified.

4. *Illustrations of finding Moments of Inertia by Integration.*

(*a*) A uniform rod of small cross-section about a line perpendicular to it at one end.

Here, if the length of the rod be $2\,a$, and the density ρ,

$$I = \Sigma \rho dx \cdot x^2 = \int_0^{2a} \rho x^2 dx = M \frac{4\,a^2}{3}.$$

(*b*) A circular arc of uniform density about an axis through
its midpoint perpendicular to its plane.

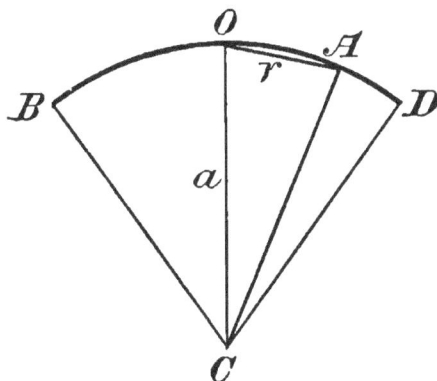

Fig. 1.

In Fig. 1, let $OA = r$, $OCA = \theta$, $OCB = \alpha$; then the moment
of inertia of the arc BOD about an axis through O perpen-
dicular to the plane of the paper is $2\,\Sigma \rho ds \cdot r^2$, where ds is an
element of the arc at A.

$$\therefore \ I = 8\,\rho a^3 \int \sin^2\frac{\theta}{2}\,d\theta = 4\,\rho a^3 \int_0^a (1 - \cos\theta)\,d\theta = 2\,M\left(1 - \frac{\sin\alpha}{\alpha}\right)a^2.$$

(c) An elliptic plate, of small thickness and uniform density.

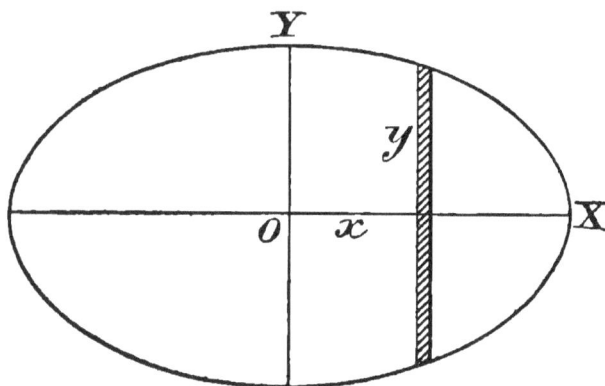

Fig. 2.

In Fig. 2, divide the plate into strips, and then we have

$$I \text{ about } OY = 4\int_0^a \rho x^2 y\,dx = 4\,\rho\int_0^a \frac{b}{a}x^2\sqrt{a^2 - x^2}\,dx = M\frac{a^2}{4}.$$

Similarly, I about $OX = M\dfrac{b^2}{4}.$

And I about a line through O perpendicular to the plate will evidently be $M\dfrac{a^2 + b^2}{4}.$

For a circular plate $a = b$.

(d) A rectangular plate, sides $2\,a$, $2\,b$.

By dividing the plate into strips of mass m it will be seen that

$$I \text{ about side } 2\,a = \Sigma\left(m\,\frac{4}{3}\frac{b^2}{3}\right) = M\frac{4}{3}\frac{b^2}{3},$$

and $\qquad I \text{ about side } 2\,b = \Sigma\left(m\,\frac{4}{3}\frac{a^2}{3}\right) = M\frac{4}{3}\frac{a^2}{3}.$

Also, I about a line through a corner perpendicular to the plate is $M\frac{4}{3}(a^2 + b^2)$. For a square plate $a = b$.

(*e*) A triangular plate.

Let the triangle be ABC, and choosing C as origin of coördinates, let CA, CB be the axes.   Then, dividing the triangle into strips parallel to AC, an elemental mass at (x, y) is equal to $\rho dxdy \sin C$, ρ being the density.   The distances of this element from AC, BC, and the point C are $x \sin C$, $y \sin C$, and $\sqrt{x^2+y^2+2xy\cos C}$.

Hence I about $AC = \displaystyle\int_0^a \int_0^{\frac{b}{a}(a-x)} \rho x^2 \sin^3 C dxdy,$

I about $BC = \displaystyle\int_0^a \int_0^{\frac{b}{a}(a-x)} \rho y^2 \sin^3 C dxdy,$

and I about a line through C perpendicular to the triangle

$$= \int_0^a \int_0^{\frac{b}{a}(a-x)} \rho \sin C (x^2+y^2+2xy \cos C)dxdy.$$

These integrals can easily be evaluated, and the moments of inertia expressed in terms of the two sides and included angle.

(*f*) A sphere about a diameter.

Dividing the sphere up into small circular plates of thickness

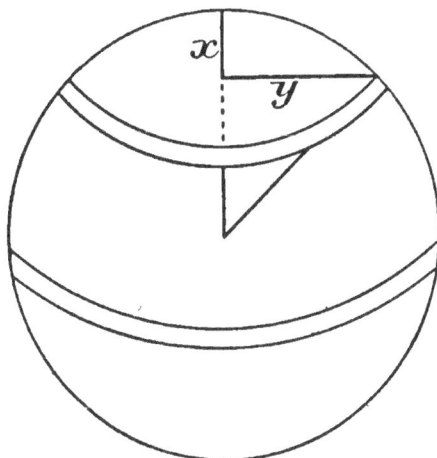

Fig. 3.

dx, as in Fig. 3, we have

I about a diameter $= 2\,\Sigma\rho \cdot \pi y^2 dx \cdot \dfrac{y^2}{2} = \rho\pi \displaystyle\int_0^a (a^2 - x^2)^2 dx = M\,\tfrac{2}{5}\,a^2.$

(g) A right circular cone, about its axis.

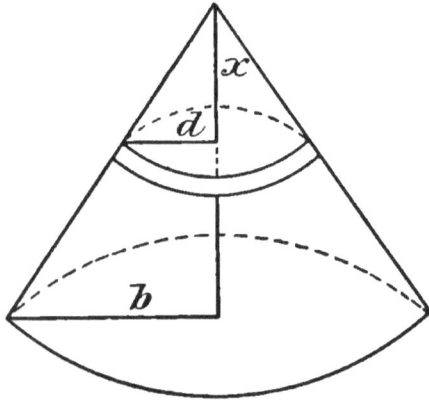

Fig. 4.

Dividing the cone up into circular strips, perpendicular to its axis, as in Fig. 4, we have, if a be its height,

$$I = \Sigma\rho\pi y^2 \cdot dx \cdot \frac{y^2}{2}, \text{ and } y = \frac{bx}{a}.$$

$$\therefore\ I = \pi\rho\,\frac{b^4}{2\,a^4}\int_0^a x^4 dx = M\frac{3}{10}\frac{b^2}{}.$$

5. Products of inertia can be evaluated in a similar way; but as they are generally eliminated from the equations of motion by a proper choice of axes, their absolute values in terms of known quantities are seldom required.

6. Although integration gives directly the values of moments and products of inertia, yet the process becomes tedious for many bodies; and the following propositions will be found useful for their determination, when one knows the position of the centre of inertia.

6 RIGID DYNAMICS.

Proposition I. — *To connect moments and products of inertia of a rigid body about any axes with moments and products of inertia about parallel axes through the centre of inertia.*

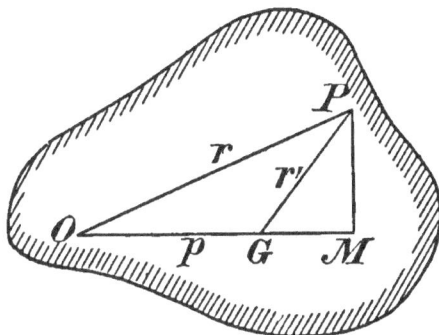

Fig. 5.

Let the plane of Fig. 5 represent any plane of the body perpendicular to the two parallel axes, of which one cuts this plane in the point O, and the other passing through the centre of inertia cuts it in G. Then for any point P in this plane, we have

$$r^2 = p^2 + r'^2 + 2p \cdot GM.$$

Hence for the whole body we must have

$$\Sigma mr^2 = \Sigma m(p^2 + r'^2 + 2p \cdot GM)$$
$$= \Sigma mp^2 + \Sigma mr'^2 + 2p \cdot \Sigma mGM$$
$$= Mp^2 + \Sigma mr'^2, \quad \text{since } \Sigma mGM = 0.$$

Or, as it may be written

$$I = I_G + Mp^2,$$

where I is the moment of inertia about any axis, and I_G is that about a parallel axis through the centre of inertia, and p is the perpendicular distance between the axes.

If three parallel axes be taken in a body, of which the third passes through the centre of inertia, and a plane be taken cutting these axes perpendicularly at the points O, O', G, then we can prove for the whole body, as before, that

$$I \text{ about axis through } O = I^G + Ma^2,$$

$$I' \text{ about axis through } O' = I_G + Mb^2,$$

and $$\therefore I = I' + M(a^2 - b^2),$$

where $$OG = a, \quad O'G = b.$$

If G happens to be in the line OO', this relation is much sim-plified. Also, if $O'G$ is at right angles to OO', then

$$I = I' + M(OO')^2.$$

which relation is sometimes useful in the case of symmetrical bodies.

It is evident, moreover, from these relations that, of all straight lines having a given direction in a body, the least moment of inertia is about that one which passes through the centre of inertia.

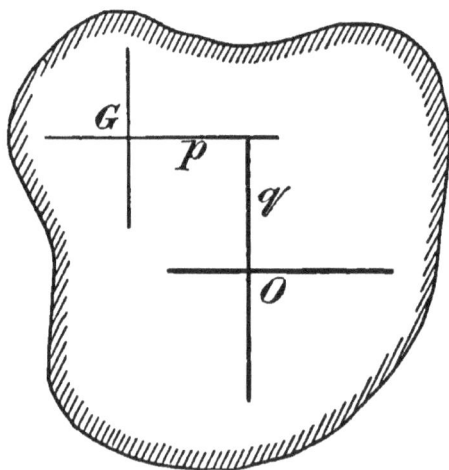

Fig. 6.

In the case of products of inertia, similar results may be obtained. Thus, if we require the product of inertia with regard

to any two coördinate planes of a body, let parallel planes be
taken passing through the centre of inertia. Let the plane of
the paper in Fig. 6 be any plane of the body perpendicular to
these four planes. Then, if P be any point whose coördinates
referred to the two sets are (x, y) and (x', y'), we must have for
the whole body

$$\Sigma mxy = \Sigma m(x'+p)(y'+q)$$
$$= \Sigma mx'y' + p\Sigma mx' + q\Sigma my' + \Sigma mpq.$$
$$\therefore \ \Sigma mxy = \Sigma mx'y' + M \cdot pq.$$

PROPOSITION II. — *In the case of a lamina, the moment of
inertia about any axis perpendicular to its plane is equal to the
sum of the moments about any two perpendicular lines drawn in
the plane through the point where the axis meets the lamina.*

For $\quad I = \Sigma m(x^2+y^2) = \Sigma mx^2 + \Sigma my^2.$

PROPOSITION III. — *To find the moment of inertia of a body
about any line, knowing the moment and products of inertia
about any three rectangular axes drawn through some point on
this line.*

In Fig. 7 let the three rectangular axes be OX, OY, OZ, and
let P be any point of the body (x, y, z), and ON any line drawn
from O, inclined at angles α, β, γ to the axes.

Then I about $ON = \Sigma mPN^2$, PN being perpendicular to ON,

and $PN^2 = OP^2 - ON^2 = (x^2+y^2+z^2) - (x\cos\alpha + y\cos\beta + z\cos\gamma)^2$

$$= (x^2+y^2+z^2)(\cos^2\alpha + \cos^2\beta + \cos^2\gamma) - (x\cos\alpha$$
$$+ y\cos\beta + z\cos\gamma)^2$$

$$= (y^2+z^2)\cos^2\alpha + \cdots + \cdots - 2yz\cos\beta\cos\gamma - \cdots - \cdots.$$

$$\therefore \ I = \Sigma m\{(y^2+z^2)\cos^2\alpha + \cdots + \cdots\} - 2\Sigma m\{yz\cos\beta\cos\gamma$$
$$+ \cdots + \cdots\}$$

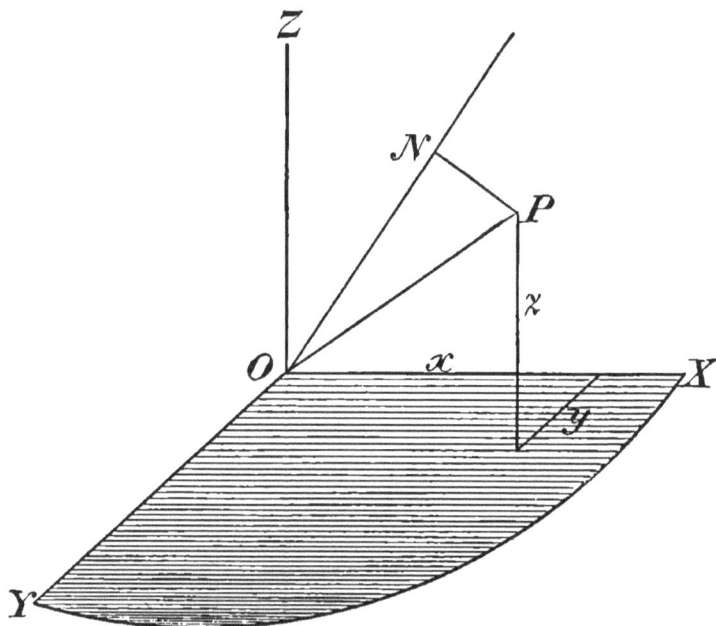

Fig. 7.

$$= A \cos^2 \alpha + B \cos^2 \beta + C \cos^2 \gamma - 2\, D \cos \beta \cos \gamma$$
$$- 2\, E \cos \gamma \cos \alpha - 2\, F \cos \alpha \cos \beta,$$

A, B, C being moments of inertia about the three axes, and D, E, F products of inertia with regard to the coördinate planes.

In this expression, it will be seen that if the axes of coördinates be so chosen that D, E, F vanish, then

$$I = A \cos^2 \alpha + B \cos^2 \beta + C \cos^2 \gamma.$$

Axes for which this holds are called *Principal Axes*, and A, B, C *Principal Moments* In many cases such axes can be found by inspection. Thus, if a body be a lamina, one principal axis at any point is the perpendicular at that point. Also, if a body be one of revolution, the axis of revolution must be a principal axis at every point of its length. And it may be stated as a general rule that axes of symmetry are principal axes.

7. In most of the problems dealing with the motion of extended bodies the axis about which the moment of inertia is to be found usually passes through the body; but it is apparent that the preceding propositions apply equally to all cases where the axes about which moments of inertia are required do not cut the body. Thus in the first proposition the axes parallel to that passing through the centre of inertia need not cut the body; in the case of a lamina, the moment of inertia about any line perpendicular to the lamina and yet not intersecting it will still be the sum of the moments about any two perpendicular lines drawn at the point where the axes meet the plane of the lamina produced; and similarly the moment of inertia about any line outside of a body will be known when we know, at any point on this line, the moments and products of inertia with respect to any three rectangular axes drawn through this point.

8. *Townsend's Theorem.*

A closed central curve, of any magnitude and form, being supposed to revolve round an arbitrary axis in its plane not intersecting its circumference; the moment of inertia with respect to the axis of revolution of the solid generated by its area is given by the formula

$$I = M(a^2 + 3 h^2),$$

where M is the mass of the solid generated, a the distance of the centre of the generating area from the axis of revolution, and h the arm length of the moment of inertia of the area with respect to a parallel axis through its centre.

For, if dA be an element of generating area,

$$I = 2 \pi \rho \cdot \Sigma[(a + x)^3 dA],$$

ρ being the density, and x a variable coördinate.

$$\therefore \ I = 2 \pi \rho \Sigma[(a^3 + 3 a^2 x + 3 a x^2 + x^3) dA].$$

But, by the symmetry of the generating area with respect to its centre, $\Sigma(xdA)=0$ and $\Sigma(x^3dA)=0$.

$$\therefore\ I = 2\,\pi\rho\Sigma[(a^3 + 3\,ax^2)dA],$$

$$\therefore\ I = M(a^2 + 3\,h^2).$$

Illustrative Examples on Moments of Inertia.

⌐ 1. Find the moment of a rectangular plate about a diagonal, the sides being $2a$, $2b$.

In this, applying Proposition III., we have

$$I = A\cos^2\theta + B\sin^2\theta,$$

the centre of the plate being the origin, and A, B principal moments.

$$\therefore\ I = M\frac{2}{3}\frac{a^2b^2}{a^2+b^2}.$$

2. A sphere or a circular plate, about a tangent. Apply Proposition I.

3. Find the moments of inertia of a rectangular parallelopiped and of a cube, about their axes of symmetry; also about a diagonal.

4. The moment of inertia of a right circular cone about a slant side is $M\dfrac{3}{20}\dfrac{b^2}{}\cdot\dfrac{6\,a^2+b^2}{a^2+b^2}$, a being the height and b the radius of the base.

5. If a is the length and b the radius of a right circular cylinder, the moment of inertia about an axis through the centre of inertia perpendicular to its axis is $\dfrac{M}{4}\left(\dfrac{a^2}{3}+b^2\right)$.

6. The moment of inertia of a pendulum bob, density ρ, in the form of an equi-convex lens of thickness $2t$ and radius a, about its axis is $\pi\rho\displaystyle\int_0^t (2\,ax - x^2)^2 dx$.

7. Find the moment of inertia of an anchor ring about its axis.

8. The moments of inertia of an ellipsoid about its three axes are $M\dfrac{b^2+c^2}{5}$, $M\dfrac{c^2+a^2}{5}$, $M\dfrac{a^2+b^2}{5}$

To find these, either divide the solid ellipsoid up into elliptic plates, or deduce from the case of a sphere.

9. A triangular plate of uniform density.

(1) To find the moment of inertia about the side BC.

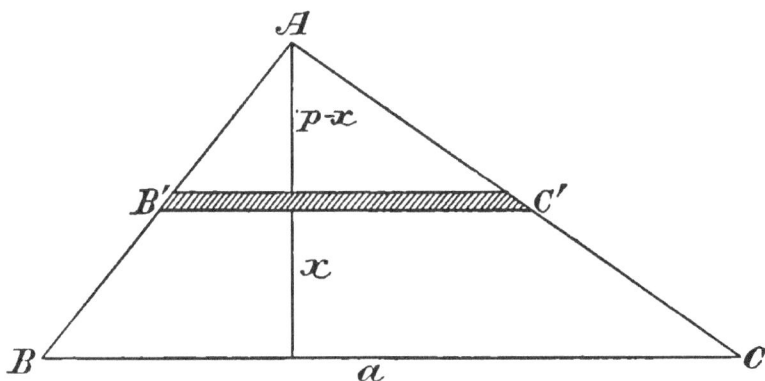

Fig. 8.

In Fig. 8, divide the triangle into strips of mass $\rho y\,dx$, where $y = B'C'$.

Then I about $BC = \Sigma \rho y\,dx \cdot x^2 = \rho \int_0^p \dfrac{p-x}{p} ax^2 dx = M \cdot \dfrac{p^2}{6}$.

(2) About a line through the centre of inertia parallel to BC.

$$I = M\left(\frac{p^2}{6} - \frac{p^2}{9}\right) = M\frac{p^2}{18}.$$

(3) About a line through A parallel to BC.

$$I = M\frac{p^2}{2}.$$

(4) About a median line.

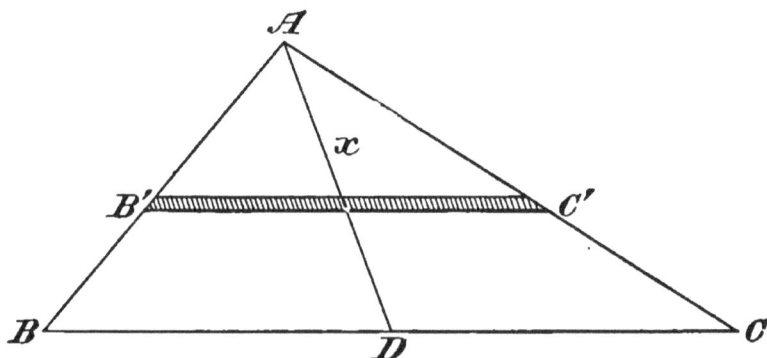

Fig. 9.

In Fig. 9, divide the triangle into strips parallel to BC, as before, and let $y = B'C'$. Then the mass of a strip is $\rho y dx \sin D$, and its moment of inertia about AD is $\rho y dx \sin D \cdot \dfrac{y^2}{12} \sin^2 D$.

Hence I of triangle about $AD = \dfrac{\rho \sin^3 D}{12} \displaystyle\int_0^{AD} y^3 dx$, and $y = \dfrac{ax}{AD}$.

$$\therefore I = M \frac{a^2 \sin^2 D}{24}.$$

(5) About a line through A, perpendicular to the plane of the triangle.

Use Fig. 9, and the moment of inertia will be found to be $\dfrac{M}{4}\left(b^2 + c^2 - \dfrac{a^2}{3}\right)$, a, b, c being the three sides.

(6) About a line through the centre of inertia, perpendicular to the plane of the triangle.

$$I = \frac{M}{36}(a^2 + b^2 + c^2).$$

10. Find the moment of inertia of a hemisphere about

(1) Its axis.

(2) A tangent at its vertex.

(3) A tangent to the circumference of its base.

(4) A diameter of its base.

11. The moment of inertia of an ellipsoidal shell of mass M about the major axis is $M\dfrac{b^2+c^2}{3}$. For a spherical shell about a diameter, $I=M\frac{2}{3}a^2$.

Deduce, by differentiation, from the ellipsoid and the sphere.

12. For an oblate spheroid (such as the earth), of excentricity ϵ, composed of similar strata of varying density, the moment of inertia about its polar axis is $\frac{8}{3}\pi\sqrt{1-\epsilon^2}\int_0^a \rho x^4 dx$, where a is the equatorial radius and ρ the density at a distance x from the centre. This can be integrated when the law of ρ is known.

13. The moment of inertia of a paraboloid of revolution about its axis of figure is $M\cdot\dfrac{r^2}{3}$, where r is the radius of the base.

14. The moment of inertia of the parabolic area cut off by any ordinate distant x from the vertex is $M\frac{3}{7}x^2$ about the tangent at the vertex, and $M\dfrac{y^2}{5}$ about the axis, where y is the ordinate corresponding to x.

15. The radius of gyration of a lamina bounded by the lemniscate $r^2=a^2\cos 2\theta$, (1) about its axis is $\dfrac{a}{4}\sqrt{\pi-\frac{8}{3}}$; (2) about a line in the plane of the lamina through the node and perpendicular to the axis is $\dfrac{a}{4}\sqrt{\pi+\frac{8}{3}}$; (3) about a tangent at the node is $\dfrac{a}{4}\sqrt{\pi}$.

16. To find the radius of gyration of a lamina bounded by a parallelogram about an axis perpendicular to it through its centre of inertia. (Euler.)

If $2a$, $2b$, be the lengths of two adjacent sides of the parallelogram, then, whatever be their inclination,

$$K^2=\frac{a^2+b^2}{3}.$$

17. To find the radius of gyration of a hollow sphere about a diameter. (Euler.)

$$K^2 = \frac{2}{5} \cdot \frac{a^5 - b^5}{a^3 - b^3},$$

a and b being the external and internal radii.

18. To find the radius of gyration of a truncated cone about its axis. (Euler.)

$$K^2 = \frac{3}{10} \cdot \frac{a^5 - b^5}{a^3 - b^3},$$

a, b, being the radii of its ends.

19. The moment of inertia of a lamina bounded by a regular polygon of n sides, each of length $2a$, about an axis through its centre perpendicular to its plane is

$$\frac{Ma^2}{6}\left(1 + 3\cot^2\frac{\pi}{n}\right).$$

And from this it can be seen that the moment of inertia about any line in the plane of the lamina through the centre

$$\frac{Ma^2}{12}\left(1 + 3\cot^2\frac{\pi}{n}\right).$$

20. A quantity of matter is distributed over the surface of a sphere of radius a, so that the density at any point varies inversely as the cube of the distance from a point inside distant b from the centre. Find the moment of inertia about that diameter which passes through the point inside, and prove that the sum of the principal moments there is equal to $2M(a^2 - b^2)$.

What if the point be outside?

CHAPTER II.

ELLIPSOIDS OF INERTIA AND PRINCIPAL AXES.

9. *Ellipsoids of Inertia.*

At any point O in a rigid body let there be taken three rectangular axes OX, OY, OZ, as in Fig. 10. Describe with O as centre the ellipsoid,

$$Ax^2 + By^2 + Cz^2 - 2\,Dyz - 2\,Ezx - 2\,Fxy = c,$$

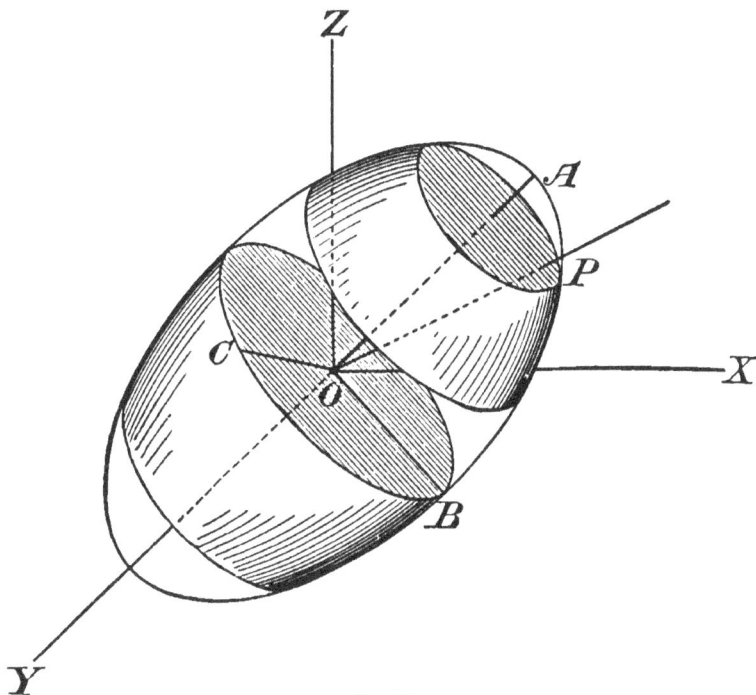

Fig. 10.

16

where A, B, C, D, E, F, have the meanings already attached to them, and are positive. Then, if OP be any line drawn from O, and cutting the ellipsoid in the point P, the moment of inertia of the body about OP is

$$A \cos^2 \alpha + B \cos^2 \beta + C \cos^2 \gamma - \cdots = I,$$

where α, β, γ are the angles which OP makes with the coördinate axes.

But if x, y, z, are the coördinates of the point P, and if $OP = r$, we must have, since the point is on the ellipsoid,

$$Ir^2 = Ax^2 + By^2 + Cz^2 - \cdots = c.$$

And since this relation is true for any position of OP, we see that the moment of inertia about any line drawn from O will be inversely proportional to the square of the corresponding radius vector cut off by the ellipsoid. Any such ellipsoid is called a *Momental Ellipsoid*.

10. If we refer the ellipsoid to its axes OA, OB, OC, then D, E, F disappear, and the axes of the ellipsoid are therefore what we have defined as *Principal Axes*.

11. It is evident that any set of principal axes at a point might be found in the foregoing manner, namely, by constructing a momental ellipsoid at the point in question and transforming to the axes of figure, which would therefore give the directions of the principal axes. And it may be stated also that three principal axes necessarily exist at each point in space for a rigid body, since the above process can always be performed.

12. From the properties of the momental ellipsoid it follows that at any point there is, in general, a line of greatest moment and also one of least moment ; if the ellipsoid degenerates into a spheroid, the moments of inertia about all diameters perpendicular to the axis of the spheroid are the same ; if it becomes a sphere, as in the case of all regular solids at their centres, the moments of inertia about all lines through the centre are

c

equal, a proposition which can be applied with advantage to the cube, proving that the moments of inertia about all lines through the centre are the same.

13. For a lamina, at any point, the section made by the corresponding momental ellipsoid is called the *Momental Ellipse* of the point.

Illustrative Examples.

1. To construct a momental ellipsoid at one of the corners of a cube.

Taking the edges as axes, $A=B=C$, $D=E=F$, and the equation for the momental ellipsoid becomes

$$A(x^2+y^2+z^2) - 2D(xy+yz+zx) = c,$$

which on transformation would give a spheroid of the form

$$A'x^2 + B'(y^2+z^2) = c',$$

and it can be seen that one principal axis is the diagonal through the corner in question, and any two lines at right angles to one another and to the diagonal will be the other two principal axes.

2. To find the momental ellipsoid at a point on the edge of a right circular cone.

Choosing axes OX, OY, OZ, as in Fig. 11, it is evident by inspection that $D=F=0$, and the axis OY is one principal axis. Then, if $AB=a$, $OB=b$; $BG=\frac{1}{4}a$, and $A=M\left(\frac{3\,b^2}{20}+\frac{a^2}{10}\right)$, $B=A+Mb^2$, $C=M\frac{13\,b^2}{10}$, $E=M\frac{ab}{4}$, and the equation of the momental ellipsoid at O is

$$(3\,b^2+2\,a^2)x^2 + (23\,b^2+2\,a^2)y^2 + 26\,b^2z^2 - 10\,abxz = c.$$

The momental ellipsoid at the point A, or at any point along the axis AB, is a spheroid.

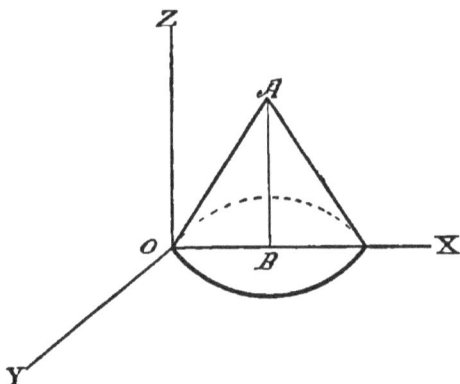

Fig. 11.

3. The momental ellipsoid at a point on the rim of a hemi-sphere is

$$2x^2 + 7(y^2 + z^2) - \tfrac{15}{4} xz = c.$$

4. The momental ellipsoid at the centre of an elliptic plate is

$$\frac{x^2}{a^2} + \frac{y^2}{b^2} + \left(\frac{1}{a^2} + \frac{1}{b^2}\right)z^2 = c.$$

5. The momental ellipsoid at the centre of a solid ellipsoid is

$$(b^2 + c^2)x^2 + (c^2 + a^2)y^2 + (a^2 + b^2)z^2 = c'.$$

14. *The Ellipsoid of Gyration.*

If at a point in a body an ellipsoid be constructed such that the moment of inertia about any perpendicular drawn from the origin on a tangent plane is equal to Mp^2, where M is the mass of the body and p the length of the perpendicular, it is called an *ellipsoid of gyration.* And, since, referred to its axis, we have by definition $A = Ma^2$ about the axis of x, $B = Mb^2$ about the axis of y, and $C = Mc^2$ about the axis of z, its equation must be

$$\frac{x^2}{A} + \frac{y^2}{B} + \frac{z^2}{C} = \frac{1}{M}.$$

This ellipsoid may also be used to indicate the directions of the principal axes; and, from the form of its equation, it is apparent that it is co-axial, but not similarly situated, with a momental ellipsoid.

15. When ellipsoids are constructed at the centres of inertia, it is customary to speak of them as *central ellipsoids*.

16. *Equimomental Systems.*

Two systems are equimomental when their moments of inertia about all lines in space are equal each to each. And from this definition, taken along with the two fundamental propositions already proved, —

$$I = I_G + Mp^2,$$

$$I = A \cos^2 \alpha + B \cos^2 \beta + C \cos^2 \gamma,$$

it follows that systems will be equimomental when they have

1. The same mass and centre of inertia.
2. The same principal axes at the centre of inertia.
3. The same principal moments at the centre of inertia.

In some particular cases we may, instead of considering a system or single body, use a simple equimomental system in determining its motion; but generally the labour of proving that systems are equimomental, or of finding a simple system which will be equimomental with a complicated one, is greater than that of solving the problem directly. The following examples, however, will serve to show how the process is carried out.

Illustrative Examples.

1. Show that three masses, each equal to $\dfrac{M}{3}$, placed at the middle points of the sides of a triangular plate of mass M, are equimomental with the triangle.

If this equimomental system be assumed, all the problems in connection with a triangular plate, such, for example, as finding moments of inertia about the sides, perpendiculars, and median lines, are very much simplified; but the difficulty of proving this assumption is greater than that of solving the problems, as has already been done by a direct process.

2. In an elliptic plate, find three points on the boundary at which, if three masses each equal to $\dfrac{M}{3}$ be placed, they will form a system equimomental with the plate, whose mass is M.

3. Show that three points can always be found in a plane area of mass M, so that three masses, each equal to $\dfrac{M}{3}$, placed at these points will form a system equimomental with the area.

The situation of the points is shown in Fig. 12, which represents the momental ellipse at the centre of inertia of the area. A may be anywhere on the boundary of the ellipse; B and C are so situated that $BD=DC$ and $OD=DE$.

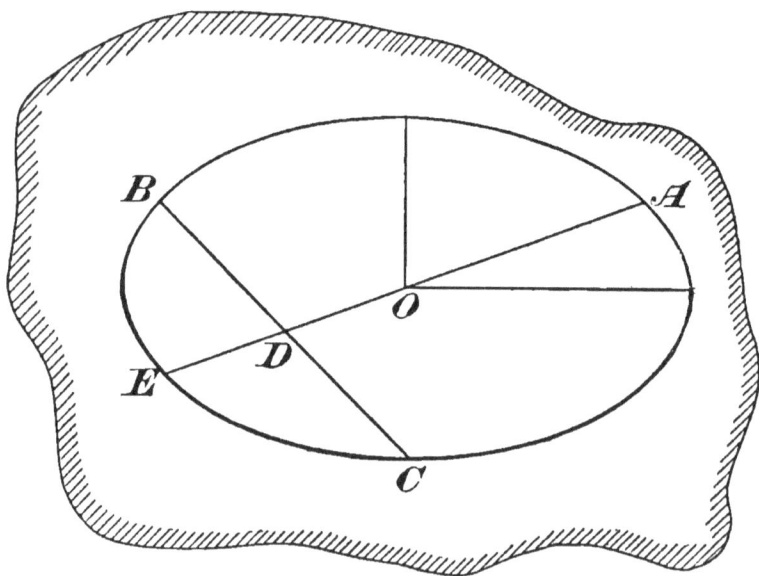

Fig. 12.

4. Find the momental ellipse at the centre of gravity of a triangular area.

5. The momental ellipse at an angular point of a triangular area touches the opposite side at its middle point, and bisects the adjacent sides.

17. *Principal Axes.*

To find the principal axes at any point of a rigid body, three rectangular axes might be chosen, and the conditions $\Sigma mxy = 0$, $\Sigma myz = 0$, $\Sigma mzx = 0$, would be sufficient to solve the problem, either by direct analysis or by the construction and subsequent transformation of the equation of the momental ellipsoid. But this process would often be tedious, and is generally unnecessary. Usually, by inspection, one at least of the principal axes can be found, as has been already mentioned, and then the other two may be obtained by the following propositions.

Given one principal axis at a point, to find the other two.

Let O be any point in the body, and let OZ, drawn perpendicular to the plane of the paper be one perpendicular axis. Take any two lines, OX, OY, at right angles to one another as

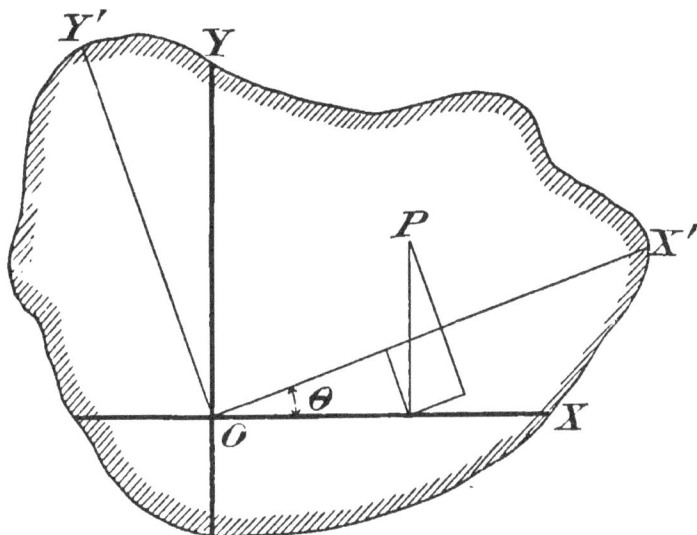

Fig. 13.

axes in this plane, and let OX', OY' be the other two principal axes at O.

Then if P be any point (x, y) or $(x'y')$, and the body extends above and below the plane of the paper, we must have as a condition that OX', OY' shall be principal axes, $\Sigma mx'y' = 0$ throughout the body. But $x' = x \cos \theta + y \sin \theta$, and $y' = -x \sin \theta + y \cos \theta$.

Therefore the condition becomes

$$\Sigma m\{-x^2 \sin \theta \cos \theta + y^2 \sin \theta \cos \theta + xy \overline{\cos^2 \theta - \sin^2 \theta}\} = 0,$$

which becomes, on reduction,

$$\tan 2\theta = \frac{2 \Sigma mxy}{\Sigma mx^2 - \Sigma my^2} = \frac{2F}{B - A},$$

according to our previous notation.

If, then, A, B, F be found in respect of any two rectangular axes OX, OY, θ is known, and therefore the position of OX', OY'.

18. The condition that a line shall be a principal axis at some point of its length is, that taking the line as axis of z and the point as origin, the relations $\Sigma mxz = 0$, $\Sigma myz = 0$ shall be satisfied. It is not true, however, that if a line be a principal axis at one point of its length, it will be a principal axis at any other, or at all points of its length. For example, in Fig. 11, the line OX is a principal axis at the point B on account of the symmetry of the cone, but it is not a principal axis at the point O. Similarly, in a hemisphere, any diameter of the base is a principal axis at the centre of the base, but not at a point on the rim. There is one case, however, in which a line is a principal axis throughout its length, and as this is of some importance, the following statement and simple proof are given.

19. *If a line be a principal axis at the centre of inertia, it will be a principal axis at every point of its length.*

Let a portion of the body be represented in Fig. 14, O being
the centre of inertia, OO' the principal axis at the centre of
inertia, OX, OY any rectangular axes at O, perpendicular to
OO', and OX', OY' parallel axes through O'.

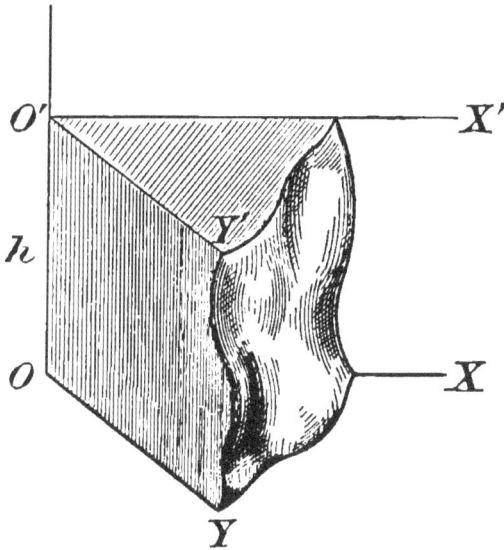

Fig. 14.

Then we have, by a previous proposition:

$$\Sigma mx'z' \text{ at } O' = \Sigma mxz \text{ at } O + M(h\bar{x}),$$

and $\qquad \Sigma my'z' \text{ at } O' = \Sigma myz \text{ at } O + M(h\bar{y}).$

But $\qquad \bar{x} = \bar{y} = 0 = \Sigma mxz = \Sigma myz$, by hypothesis.

$\therefore$ at O' $\qquad \Sigma mx'y' = 0 = \Sigma my'z'$,

and therefore OO' is a principal axis at O', and therefore also
at any point in its length.   Conversely, it may be shown that
if a line be a principal axis at all points in its length, it must
pass through the centre of inertia.

20. To determine the locus of points at which the momental ellipsoid for a given body degenerates to a spheroid, and the points, if such exist, at which it becomes a sphere.

Let the body be referred to its principal axes at the centre of inertia, and let A, B, and C be its principal moments, and M its mass : —

(1) If all three moments be unequal, say $A > B > C$, there will be no point at which the momental ellipsoid for that body will be a sphere, but at any point P on the ellipse

$$\frac{x^2}{A-C} + \frac{y^2}{B-C} = \frac{I}{M}, \quad z = 0,$$

or on the hyperbola,

$$\frac{x^2}{A-B} - \frac{z^2}{B-C} = \frac{I}{M}, \quad y = 0,$$

it will be a spheroid with axes of revolution touching the conic at P. The momental ellipsoid at all other points will have three unequal axes.

(2) If two of the moments be equal, and each less than the third, say $A > B = C$, there will be two points at which the momental ellipsoid for that body will be a sphere, viz., the points on the axis of x, distant $\pm\sqrt{\left(\frac{A-C}{M}\right)}$ from the centre of inertia. At every other point on the axis of x, the momental ellipsoid will be a spheroid with the axis of x as axis of revolution. At all points not on the axis of x the momental ellipsoid will have three unequal axes.

(3) If two of the moments be equal, and each greater than the third, say $A = B > C$, the momental ellipsoid for that body will be a spheroid at every point on the axis of z, or on the circle,

$$x^2 + y^2 = \frac{A-C}{M}, \quad z = 0.$$

At all other points it will have three unequal axes.

(4) If $A = B = C$, the momental ellipsoid will be a sphere at the centre of inertia and a spheroid at every other point.

From the above it is seen at once that in the majority of bodies there is no point for which all axes drawn through it are principal axes.

Illustrative Examples.

1. To find the principal axes of a triangular lamina, at an angular point.

One principal axis is the line drawn through the angular point perpendicular to the lamina, and the other two are found in the following way. In Fig. 15, let OA, OB be two rectangular axes, OX, OY the principal axes in the plane of the lamina. Then the angle which OX makes with OA will be given by the formula $\tan 2\theta = \dfrac{2F}{B-A}$, where

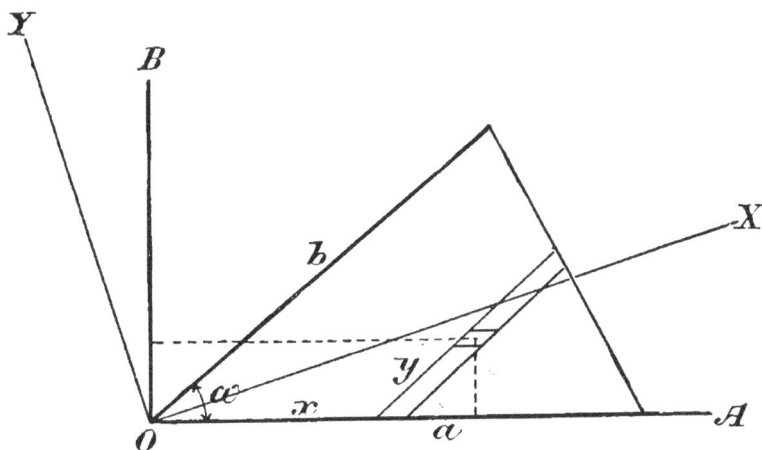

Fig. 15.

$A =$ moment of inertia about OA

$$= \int_0^a \int_0^{\frac{b}{a}(a-x)} \rho \sin^3 \omega y^2 dx dy,$$

B = moment of inertia about OB

$$= \int_0^a \int_0^{\frac{b}{a}(a-x)} \rho \sin \omega \, (x + y \cos \omega)^2 dx dy,$$

and

$$F = \int_0^a \int_0^{\frac{b}{a}(a-x)} \rho \sin \omega (x + y \cos \omega) y \sin \omega dx dy,$$

ρ being the density of the lamina, and the axes of x and y lying along the sides of the triangle.

It will be found, on evaluating these integrals, that

$$\tan 2\theta = \frac{b \sin \omega \, (a + 2 b \cos \omega)}{a^2 + ab \cos \omega + b^2 \cos 2\omega}.$$

As a simple case, let $\omega = \dfrac{\pi}{2}$; then the triangle is right angled, and $\tan 2\theta = \dfrac{ab}{a^2 - b^2}$, as can easily be found independently of the above formula.

2. To find the principal axes at any point of an elliptic lamina.

In Fig. 16, let O' be the point (a, β) at which we require the

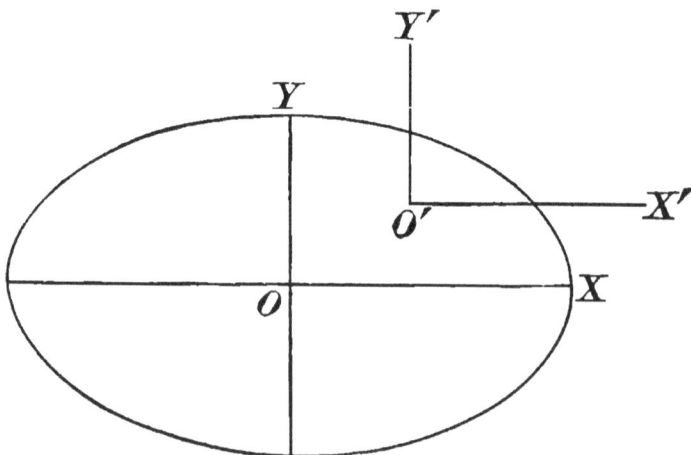

Fig. 16.

principal axes. Then the angle θ which $O'X'$ makes with the principal axis at O' is given by

$$\tan 2\,\theta = \frac{2\,F}{B-A},$$

where $A = I$ about $O'X' = I$ about $OX + M\beta^2$

$$= M\!\left(\frac{b^2}{4} + \beta^2\right),$$

$B = I$ about $O'Y' = I$ about $OY + M\alpha^2$

$$= M\!\left(\frac{a^2}{4} + \alpha^2\right),$$

and $F = \Sigma m x' y' = \Sigma m x y + M\alpha\beta = M\alpha\beta.$

$$\therefore\ \tan 2\,\theta = \frac{2\,M\alpha\beta}{M\!\left(\dfrac{a^2}{4} + \alpha^2\right) - M\!\left(\dfrac{b^2}{4} + \beta^2\right)}$$

$$= \frac{8\,\alpha\beta}{(a^2 - b^2) + 4(\alpha^2 - \beta^2)}.$$

The third principal axis is, of course, at right angles to the lamina, through the point O'.

3. To find at what point a side of a triangle is a principal axis.

Fig. 17 shows the construction and proof. BC is the side in question, and is bisected at E. AD is drawn perpendicular to BC, and DE is bisected at O. Then, taking the equimomental system $\dfrac{M}{3}$ at the middle points of the sides, in order that the inertia-product F may vanish, the principal axis perpendicular to the side BC must bisect the join of the mid-points of the sides AB and AC, hence BC, OY are the principal axes in the plane of the lamina at the point O.

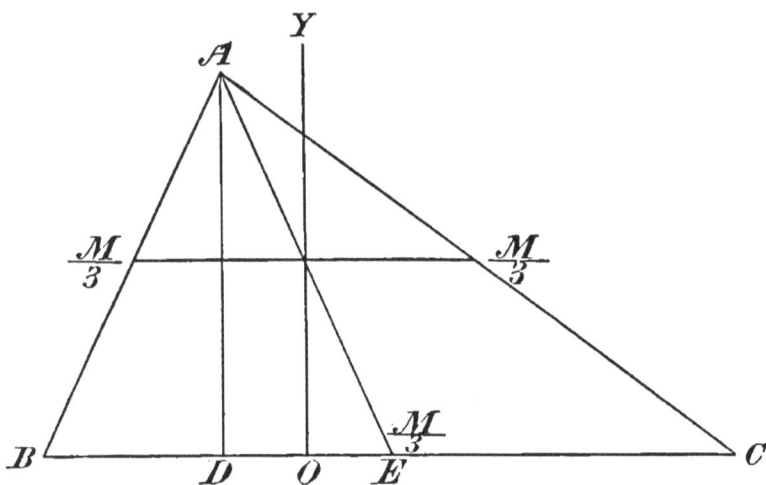

Fig. 17.

4. Find the principal axes at any point of a square or a rectangular plate.

5. Find the principal axes at any point within a cube or a rectangular parallelopiped.

6. The principal axes at any point on the edge of a hemisphere are, one touching the circumference of its base, and two others, given by the relation $\tan 2\,\theta = \frac{3}{4}$.

7. The principal axes at any point on the edge of a right circular cone are, one touching the circumference of its base, and two others, given by the relation $\tan 2\,\theta = \dfrac{10\,ab}{23\,b^2 - 2\,a^2}$, where a is the altitude of the cone and b the radius of the base.

If $a = 2\,b$, then one of the principal axes passes through the centre of inertia, and at the centre of inertia itself all axes are principal axes.

8. Find the principal axes at any point in a lamina in the form of a quadrant of an ellipse.

9. Determine the condition that the edge of any tetrahedron may be a principal axis at some point of its length, and find the point.

10. Two points P and Q are so situated that a principal axis at P intersects a principal axis at Q. Then if two planes be drawn at P and Q perpendicular to these principal axes, their intersection will be a principal axis at the point where it is cut by the plane containing the principal axes at P and Q.

(Townsend.)

21. In determining the directions of the principal axes by aid of the relation $\tan 2\theta = \dfrac{2F}{B-A}$, if $F=0$ and at the same time $B=A$, then the value of θ is indeterminate, and any two axes perpendicular to the given one and to one another are principal axes ; if $B=A$ and F is finite, then $\tan 2\theta =$ infinite and $2\theta = \dfrac{\pi}{2}$; if $F=0$, and B is not equal to A, then $\tan 2\theta = 0$ and $\theta = 0$ or $\dfrac{\pi}{2}$.

CHAPTER III.

D'ALEMBERT'S PRINCIPLE.

22. In determining the motion of a single particle of mass m, three rectangular axes are chosen, and if X, Y, Z be the accelerations in the directions of these three axes, the equation of the path is found from the relations

$$m \frac{d^2x}{dt^2} = mX,$$

$$m \frac{d^2y}{dt^2} = mY,$$

$$m \frac{d^2z}{dt^2} = mZ.$$

In the case of a body which is composed of a number of particles collected into what is termed a *rigid body*, if we follow the above method, we get three relations of the type

$$m \frac{d^2x}{dt^2} = mX + f_1,$$

where f_1 arises from the internal molecular actions, and X is, as before, the acceleration of a single particle whose mass is m. For every particle we should get similar relations, the value of f_1, however, changing from point to point in the body. We can proceed no further in the solution of such equations, owing to our imperfect knowledge of the value and variation of f_1. But the Principle of D'Alembert enables us to form an equation independent of the internal molecular actions by taking the

31

sum of all the forces acting on the individual particles which compose the body. Thus, for all the particles, we must have

$$\Sigma\left(m\frac{d^2x}{dt^2}\right)=\Sigma(mX)+\Sigma(f_1),$$

$$\Sigma\left(m\frac{d^2y}{dt^2}\right)=\Sigma(mY)+\Sigma(f_2),$$

$$\Sigma\left(m\frac{d^2z}{dt^2}\right)=\Sigma(mZ)+\Sigma(f_3),$$

and D'Alembert's principle states that

$$\Sigma(f_1)=\Sigma(f_2)=\Sigma(f_3)=0.$$

23. The equations of motion of a rigid body, then, are

$$\left.\begin{aligned}\Sigma\left(m\frac{d^2x}{dt^2}\right)&=\Sigma(mX),\\ \Sigma\left(m\frac{d^2y}{dt^2}\right)&=\Sigma(mY),\\ \Sigma\left(m\frac{d^2z}{dt^2}\right)&=\Sigma(mZ).\end{aligned}\right\}\qquad(A)$$

Each force of the type $m\frac{d^2x}{dt^2}$ is termed an *effective force;* and the above relations are equivalent to saying that the effective forces, if reversed, would be in equilibrium with the external or impressed forces ; they may be looked upon either as equations of motion or as conditions for equilibrium.

24. It is evident, also, from this same principle, that if we take the *sum* of all the moments of the effective forces, these, if reversed, will balance the sum of all the moments of the external forces. Consequently, for any set of rectangular axes, we must also have

$$\Sigma\left[m\left(y\frac{d^2z}{dt^2}-z\frac{d^2y}{dt^2}\right)\right]=L,$$

$$\Sigma\left[m\left(z\frac{d^2x}{dt^2}-x\frac{d^2z}{dt^2}\right)\right]=M,$$

$$\Sigma\left[m\left(x\frac{d^2y}{dt^2}-y\frac{d^2x}{dt^2}\right)\right]=N,$$

(B)

where L, M, N are the couples produced by the external forces.

25. It may be stated here that D'Alembert's principle holds also in the case of a system of bodies moving under their mutual actions and reactions, and applies to the motion of liquids. It is a direct consequence of Newton's Third Law of Motion.

26. *Deductions from D'Alembert's Principle.*

Taking any one of the equations (A), we have

$$\Sigma m\frac{d^2x}{dt^2}=\Sigma mX.$$

But, by definition of the centre of inertia,

$$\Sigma mx=M\bar{x},$$

and
$$\therefore\ \Sigma m\frac{d^2x}{dt^2}=M\frac{d^2\bar{x}}{dt^2}$$

Therefore the above relation becomes

$$M\frac{d^2\bar{x}}{dt^2}=\Sigma mX,$$

and similarly for the other two.

(1) *Hence, the motion of the centre of gravity of a system under the action of any forces is the same as if all the mass were collected at the centre of inertia and all the forces were applied there parallel to their former direction.*

And so the problem of finding the motion of the centre of inertia of a system, however complex, is reduced to finding that of a single particle.

D

Moreover, taking one of the equations (B),

$$\Sigma m\left(y\frac{d^2z}{dt^2} - z\frac{d^2y}{dt^2}\right) = L,$$

since we may choose the origin of coördinates at any point, let it be so chosen that at the time of forming these equations the centre of inertia is coincident with it, but moving with a certain velocity and acceleration. Then, evidently, we must obtain a relation of the same form as the foregoing, just as if we had considered the centre of inertia as a fixed point. In other words, such a relation as the above will hold at each instant of the body's motion, independently of the origin and of the position of the body.

(2) *Hence, the motion of a body, under the action of any finite forces, about its centre of inertia, is the same as if the centre of inertia were fixed and the same forces were acting on the body.*

27. The two previous deductions are known as the principles of the *Conservation of the motions of Translation and Rotation*, and show us that we may consider the two motions independently of one another.

28. *Impulsive Equations of Motion.*

Since an impulse can be measured only by the change of momentum it induces in a body, in applying D'Alembert's Principle to impulsive forces we must alter the expressions for the effective forces, which will be represented not by the products of masses and accelerations, but by the products of masses and changes of velocity. All the preceding relations will hold equally for impulsive forces if we then write changes of velocity for accelerations.

Thus, such a relation as

$$\Sigma m\frac{d^2x}{dt^2} = \Sigma mX,$$

for finite forces will become

$$\Sigma m \left\{ \left(\frac{dx}{dt}\right)' - \left(\frac{dx}{dt}\right) \right\} = \Sigma X,$$

for impulsive forces where the velocity of each particle of mass m is changed from $\frac{dx}{dt}$ abruptly to $\left(\frac{dx}{dt}\right)'$ by the action of an impulse X. And it may be said, generally, that equations of motion for impulsive forces can be obtained from the corresponding equations for finite forces by substituting in the latter changes of velocities for accelerations.

29. In forming any relations for impulses, it must be borne in mind that all finite actions, such as that of gravity, are to be neglected; after the impulse has acted, the subsequent motion will, of course, be found by applying the equations for the finite forces which usually are called into play after the impulse has operated.

Illustrative Examples.

1. A rough uniform board, of length $2a$ and mass m, rests on a smooth horizontal plane. A man of mass M walks from one end to the other. Determine the motion.

This example furnishes an excellent illustration of the truth of D'Alembert's principle, which asserts that the motion of the centre of inertia of the *system* will be the same as if we applied there all the forces *external* to the system, each acting in its proper direction. All the forces at the centre of inertia are then downwards, and as the centre of inertia cannot move downwards, it must therefore be at rest; and as the man walks along the whole board, he will therefore advance relatively to the fixed horizontal plane through a distance $\frac{2ma}{M+m}$, and the board will recede through a distance $\frac{2Ma}{M+m}$.

Analytically, we have for the motion in a horizontal direction, since there are no horizontal forces *external* to the system, the equation

$$\Sigma m \frac{d^2 x}{dt^2} = 0.$$

$$\therefore M \frac{d^2 \bar{x}}{dt^2} = 0,$$

and
$$\therefore \frac{d\bar{x}}{dt} = 0 \text{ or constant.}$$

If the man and board start from rest, as we have supposed, then

$$\frac{d\bar{x}}{dt} = 0.$$

$$\therefore \bar{x} = \text{constant,}$$

which means that the position of the centre of inertia remains unaltered throughout the motion of the two parts of the system.

2. Two persons, A and B, are situated on a smooth horizontal plane at a distance a from each other. If A throws a ball to B, which reaches B after a time t, show that A will begin to slide along the plane with a velocity $\frac{ma}{Mt}$, where M is his own mass and m that of the ball.

3. A person is placed on a perfectly smooth surface. How may he get off?

4. Explain how a person sitting on a chair is able to move the chair along the ground by a series of jerks without touching the ground with his feet.

5. How is a person able to increase his amplitude in swinging without touching the ground with his feet?

6. Explain dynamically the method of high jumping with a pole; and show that a man should be able to jump as far on a horizontal plane without a pole as with one.

7. Two coins, a large and a small one, are spun together on an ordinary table about an axis nearly vertical. Which will come to rest first, and why?

8. A circular board is placed on a smooth horizontal plane, and a dog runs with uniform speed around on the board close to its edge. Find the motion of the centre of the board.

30. *The Principle of Energy.*

Before entering upon the discussion of the motion of a rigid body, what is known as the principle of energy will be explained, as it is exceedingly useful, and often gives a partial solution of a problem without any reference to the equations of motion, and in many cases furnishes solutions which are both simple and elegant when compared with those obtained by the use of Cartesian coördinates.

If a single particle of mass m be moving along the axis of x, under the action of a force F in the same direction, we have, as the equation of motion,

$$m\frac{d^2x}{dt^2} = F.$$

And multiplying both sides by $\frac{dx}{dt}$ and integrating, we get

$$\tfrac{1}{2}m(v^2 - V^2) = \int_0^x F\,dx,$$

where V is the initial value of v or $\frac{dx}{dt}$.

The expression on the left-hand side of the equation is the change in *kinetic energy*, which is equal to the *work done* by the force from o to x.

What is true of a single force acting in a definite direction and of a single particle of mass m is also true of a number of forces acting on a rigid body or on a system. Then the analytical expression for the work done by a system of forces becomes

$$\Sigma m \int (X dx + Y dy + Z dz),$$

which must be equal to

$$\tfrac{1}{2} \Sigma m v^2 - \tfrac{1}{2} \Sigma m V^2.$$

In the general case, where bodies move with both translation and rotation, the total kinetic energy can easily be shown to be that due to translation of the whole mass collected at the centre of inertia, together with that due to rotation about the centre of inertia considered as a fixed point.

For if x, y, z be the coördinates of any particle of mass m and velocity v at time t, and $\bar{x}, \bar{y}, \bar{z}$ be the coördinates of the centre of inertia, ξ, η, ζ the coördinates of the particle referred to the centre of inertia, then the total kinetic energy is equal to

$$\tfrac{1}{2} \Sigma m v^2 = \tfrac{1}{2} \Sigma m \left\{ \left(\frac{dx}{dt}\right)^2 + \left(\frac{dy}{dt}\right)^2 + \left(\frac{dz}{dt}\right)^2 \right\}$$

$$= \tfrac{1}{2} \Sigma m \left\{ \left(\frac{d\bar{x}}{dt}\right)^2 + \left(\frac{d\bar{y}}{dt}\right)^2 + \left(\frac{d\bar{z}}{dt}\right)^2 \right\} + \tfrac{1}{2} \Sigma m \left\{ \left(\frac{d\xi}{dt}\right)^2 + \left(\frac{d\eta}{dt}\right)^2 + \left(\frac{d\zeta}{dt}\right)^2 \right\},$$

since by definition of the centre of inertia the other terms disappear. This proves the proposition.

31. According to the kind of motion and the choice of coördinates and origin, this expression for energy will assume various forms which will be given under the discussions of the special cases throughout the treatise.

Twice the energy is termed the *vis viva*.

32. To find the work done by an impulse; let Q be the measure of an impulse which, acting on a particle of mass m moving with velocity V, changes its velocity suddenly to v; then the kinetic energy is changed from $\frac{1}{2} m V^2$ to $\frac{1}{2} m v^2$.

Work done by the impulse

$$= \tfrac{1}{2} m v^2 - \tfrac{1}{2} m V^2 = \tfrac{1}{2} (mv - mV)(v + V)$$
$$= \tfrac{1}{2} Q \cdot (v + V),$$

since the impulse is measured by the change of momentum and Q is therefore equal to $mv - mV$.

A similar relation will evidently apply to a rigid body where v and V are the velocities of the point of application of the impulse resolved in the direction of the action of the impulse.

Illustrative Examples on Energy.

1. A rod OA, of length $2\,a$, fixed at O, drops from a horizontal position under the action of gravity : find its angular velocity when it is in the vertical position OB.   (See Fig. 18.)

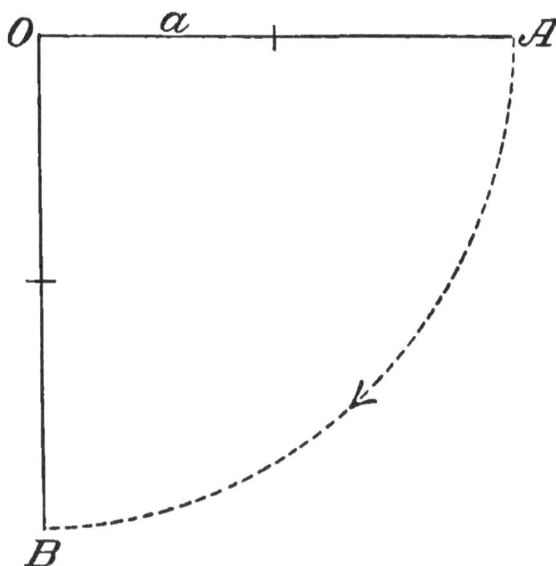

Fig. 18.

principal axes.   Then the angle θ which $O'X'$ makes with the principal axis at O' is given by

$$\tan 2\theta = \frac{2F}{B-A},$$

where $A = I$ about $O'X' = I$ about $OX + M\beta^2$

$$= M\left(\frac{b^2}{4} + \beta^2\right),$$

$B = I$ about $O'Y' = I$ about $OY + M\alpha^2$

$$= M\left(\frac{a^2}{4} + \alpha^2\right),$$

and $F = \Sigma mx'y' = \Sigma mxy + M\alpha\beta = M\alpha\beta.$

$$\therefore \tan 2\theta = \frac{2M\alpha\beta}{M\left(\frac{a^2}{4} + \alpha^2\right) - M\left(\frac{b^2}{4} + \beta^2\right)}$$

$$= \frac{8\alpha\beta}{(a^2 - b^2) + 4(\alpha^2 - \beta^2)}.$$

The third principal axis is, of course, at right angles to the lamina, through the point O'.

3. To find at what point a side of a triangle is a principal axis.

Fig. 17 shows the construction and proof. BC is the side in question, and is bisected at E. AD is drawn perpendicular to BC, and DE is bisected at O.   Then, taking the equimomental system $\frac{M}{3}$ at the middle points of the sides, in order that the inertia-product F may vanish, the principal axis perpendicular to the side BC must bisect the join of the mid-points of the sides AB and AC, hence BC, OY are the principal axes in the plane of the lamina at the point O.

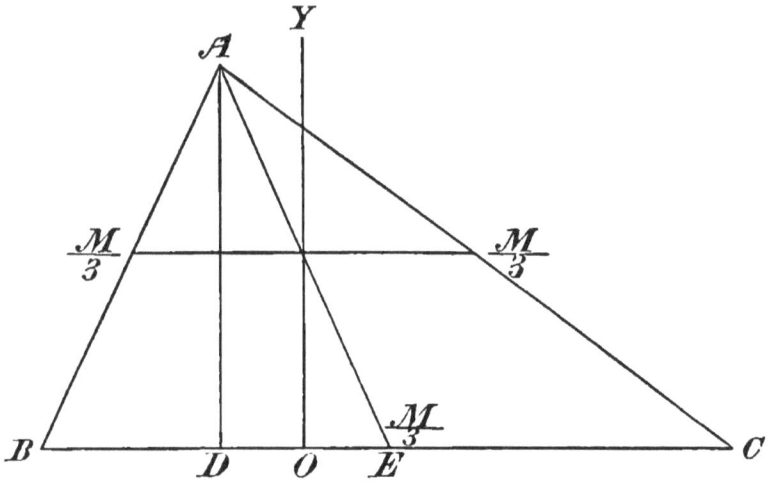

Fig. 17.

4. Find the principal axes at any point of a square or a rectangular plate.

5. Find the principal axes at any point within a cube or a rectangular parallelopiped.

6. The principal axes at any point on the edge of a hemisphere are, one touching the circumference of its base, and two others, given by the relation $\tan 2\,\theta = \frac{3}{4}$.

7. The principal axes at any point on the edge of a right circular cone are, one touching the circumference of its base, and two others, given by the relation $\tan 2\,\theta = \frac{10\,ab}{23\,b^2 - 2\,a^2}$, where a is the altitude of the cone and b the radius of the base.

If $a = 2\,b$, then one of the principal axes passes through the centre of inertia, and at the centre of inertia itself all axes are principal axes.

8. Find the principal axes at any point in a lamina in the form of a quadrant of an ellipse.

motion of translation upwards represented by $a\omega'$, and at the same time the stick keeps on rotating about the centre of inertia. Owing to the action of gravity, the motion of translation ceases, alters in direction, and finally the stick drops to the ground in an upright position. The time it takes the centre of inertia to move from its second position to its final position when the stick pitches upright is found from the well-known formula for space described under the action of gravity, which, in this case, becomes

$$a = -a\omega' \cdot t + \tfrac{1}{2}gt^2. \tag{b}$$

The condition for pitching upright is evidently to be found from the condition that the rod after leaving position (2) must rotate through $(2n+1)\dfrac{\pi}{2}$ before touching the ground, and therefore

$$\omega' \cdot t = (2n+1)\frac{\pi}{2}. \tag{c}$$

(a), (b), and (c) give the result

$$\omega^2 = \frac{g}{2a}\left(3 + \frac{p^2}{p+1}\right),$$

where
$$p = (2n+1)\frac{\pi}{2}.$$

3. A uniform heavy board hangs in a horizontal position suspended by two equal parallel strings fastened to the ends. If given a twist about a vertical axis, prove that it will rise through a distance $\dfrac{a^2\omega^2}{6g}$, where $2a$ is the length of the board, and ω the vertical twist.

4. A cannon rests on a rough horizontal plane, and is fired with such a charge that the relative velocity of the ball and cannon at the moment when the ball leaves the cannon is V. If M be the mass of the cannon, m that of the ball, and μ the coefficient of friction, show that the cannon will recoil a distance $\left(\dfrac{mV}{M+m}\right)^2 \dfrac{1}{2\mu g}$ on the plane.

5. A fine string is wound around a heavy grooved circular plate, and the free end being fixed, the plate is allowed to fall freely. Find the space described in any time.

6. A coin is spun about an axis nearly vertical upon an ordinary table. Form the equation of energy at any time as the coin descends to its position of rest.

7. A narrow, smooth, semicircular tube is fixed in a vertical plane, the vertex being at the highest point ; and a heavy flexible string, passing through it, hangs at rest. If the string be cut at one of the ends of the tube, to find the velocity which the longer portion will have attained when it is just leaving the tube.

If a be the radius of the tube, l the length of the longer portion, then, on equating the kinetic energy at the time the string is leaving the tube to the work done by gravity up to that time, it will be found that the required velocity is given by the relation

$$v^2 = ga \left\{ 2\pi - \frac{a}{l}(\pi^2 - 4) \right\}.$$

8. Explain why the grooving in a rifle barrel diminishes the force of recoil.

9. A rough wooden top in the form of a cone can rotate about its axis, which is fixed and horizontal. A fine string is fastened at the apex, and wound around it until the top is completely covered. A small weight attached to the free end is allowed to fall freely under the action of gravity, unwinding the string from the top which rotates about its axis. Find the angular velocity of the top when the string is completely unwound ; also, the equation of the path of the descending weight.

10. Two equal perfectly rough spheres are placed in unstable equilibrium, one on top of the other ; the lower sphere resting on a perfectly smooth horizontal surface. If the slightest

Let a portion of the body be represented in Fig. 14, O being the centre of inertia, OO' the principal axis at the centre of inertia, OX, OY any rectangular axes at O, perpendicular to OO', and OX', OY' parallel axes through O'.

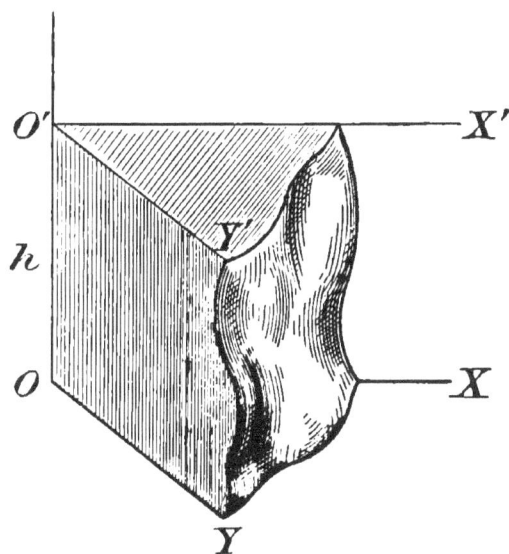

Fig. 14.

Then we have, by a previous proposition:

$$\Sigma m x'z' \text{ at } O' = \Sigma m xz \text{ at } O + M(h\bar{x}),$$

and $\qquad \Sigma m y'z' \text{ at } O' = \Sigma m yz \text{ at } O + M(h\bar{y}).$

But $\qquad \bar{x} = \bar{y} = 0 = \Sigma m xz = \Sigma m yz,$ by hypothesis.

$\therefore$ at $O' \qquad \Sigma m x'y' = 0 = \Sigma m y'z',$

and therefore OO' is a principal axis at O', and therefore also at any point in its length. Conversely, it may be shown that if a line be a principal axis at all points in its length, it must pass through the centre of inertia.

20. To determine the locus of points at which the momental ellipsoid for a given body degenerates to a spheroid, and the points, if such exist, at which it becomes a sphere.

Let the body be referred to its principal axes at the centre of inertia, and let A, B, and C be its principal moments, and M its mass : —

(1) If all three moments be unequal, say $A > B > C$, there will be no point at which the momental ellipsoid for that body will be a sphere, but at any point P on the ellipse

$$\frac{x^2}{A-C} + \frac{y^2}{B-C} = \frac{1}{M}, \quad z = 0,$$

or on the hyperbola,

$$\frac{x^2}{A-B} - \frac{z^2}{B-C} = \frac{1}{M}, \quad y = 0,$$

it will be a spheroid with axes of revolution touching the conic at P. The momental ellipsoid at all other points will have three unequal axes.

(2) If two of the moments be equal, and each less than the third, say $A > B = C$, there will be two points at which the momental ellipsoid for that body will be a sphere, viz., the points on the axis of x, distant $\pm\sqrt{\left(\dfrac{A-C}{M}\right)}$ from the centre of inertia. At every other point on the axis of x, the momental ellipsoid will be a spheroid with the axis of x as axis of revolution. At all points not on the axis of x the momental ellipsoid will have three unequal axes.

(3) If two of the moments be equal, and each greater than the third, say $A = B > C$, the momental ellipsoid for that body will be a spheroid at every point on the axis of z, or on the circle,

$$x^2 + y^2 = \frac{A-C}{M}, \quad z = 0.$$

At all other points it will have three unequal axes.

Then by D'Alembert's principle, we have

$$\Sigma m \frac{d^2x}{dt^2} = \Sigma mX + P_1 \cos \alpha_1 + P_2 \cos \alpha_2,$$

$$\Sigma m \frac{d^2y}{dt^2} = \Sigma mY + P_1 \cos \beta_1 + P_2 \cos \beta_2,$$

$$\Sigma m \frac{d^2z}{dt^2} = \Sigma mZ + P_1 \cos \gamma_1 + P_2 \cos \gamma_2,$$

X, Y, Z, being the accelerations on the unit mass m.

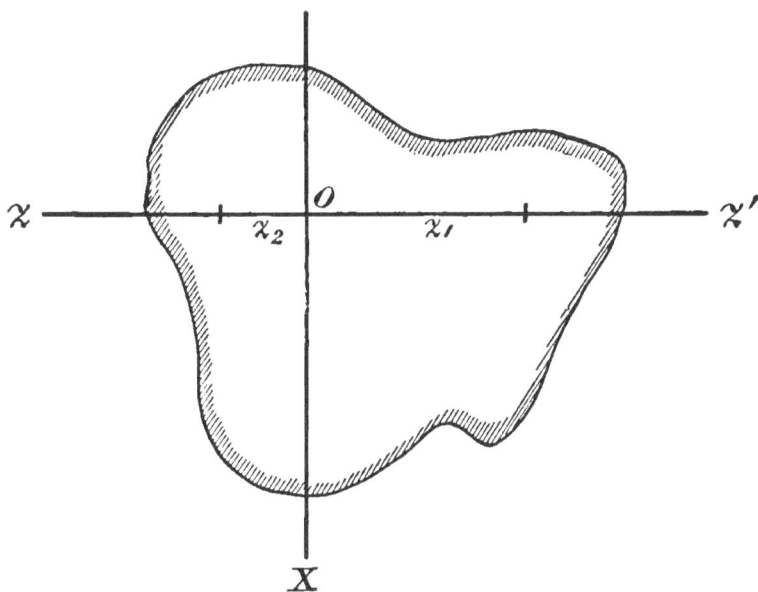

Fig. 21.

We must have, also, the relations

$$\Sigma m \left(y\frac{d^2z}{dt^2} - z\frac{d^2y}{dt^2} \right) = L \pm P_1 z_1 \cos \beta_1 \pm P_2 z_2 \cos \beta_2,$$

$$\Sigma m \left(z\frac{d^2x}{dt^2} - x\frac{d^2z}{dt^2} \right) = M \pm P_1 z_1 \cos \alpha_1 \pm P_2 z_2 \cos \alpha_2,$$

$$\Sigma m \left(x\frac{d^2y}{dt^2} - y\frac{d^2x}{dt^2} \right) = N,$$

where L, M, N, are the couples produced by the external forces.

It will be seen that there is one relation independent of the pressures

$$\Sigma m\left(x\frac{d^2y}{dt^2} - y\frac{d^2x}{dt^2}\right) = N,$$

and this gives at once, by transformation to polar coördinates,

$$\Sigma mr^2 \cdot \frac{d^2\theta}{dt^2} = N,$$

and $\therefore$ $\dfrac{d^2\theta}{dt^2} = \dfrac{\text{moment of external forces about the fixed axis}}{\text{moment of inertia about the fixed axis}},$

which, evidently, on integration gives the angular velocity at any time, and consequently the angle described in any given time.

35. *Angular Velocity of Any Heavy Body about a Fixed Horizontal Axis.*

If the body moving about a fixed horizontal axis be acted upon by gravity only, the angular velocity at any time can be

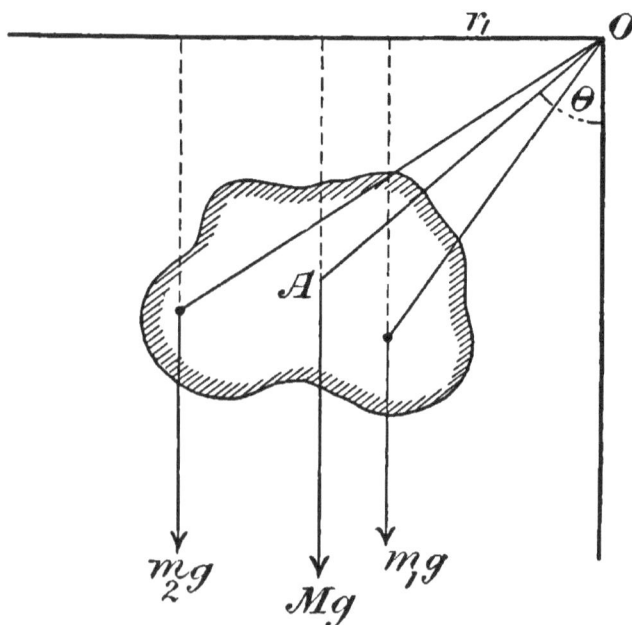

Fig. 22.

Analytically, we have for the motion in a horizontal direction, since there are no horizontal forces *external* to the system, the equation

$$\Sigma m \frac{d^2x}{dt^2} = 0.$$

$$\therefore M \frac{d^2\bar{x}}{dt^2} = 0,$$

and $$\therefore \frac{d\bar{x}}{dt} = 0 \text{ or constant.}$$

If the man and board start from rest, as we have supposed, then

$$\frac{d\bar{x}}{dt} = 0.$$

$$\therefore \bar{x} = \text{constant},$$

which means that the position of the centre of inertia remains unaltered throughout the motion of the two parts of the system.

2. Two persons, A and B, are situated on a smooth horizontal plane at a distance a from each other. If A throws a ball to B, which reaches B after a time t, show that A will begin to slide along the plane with a velocity $\frac{ma}{Mt}$, where M is his own mass and m that of the ball.

3. A person is placed on a perfectly smooth surface. How may he get off?

4. Explain how a person sitting on a chair is able to move the chair along the ground by a series of jerks without touching the ground with his feet.

5. How is a person able to increase his amplitude in swinging without touching the ground with his feet?

6. Explain dynamically the method of high jumping with a pole; and show that a man should be able to jump as far on a horizontal plane without a pole as with one.

7. Two coins, a large and a small one, are spun together on an ordinary table about an axis nearly vertical. Which will come to rest first, and why?

8. A circular board is placed on a smooth horizontal plane, and a dog runs with uniform speed around on the board close to its edge. Find the motion of the centre of the board.

30. *The Principle of Energy.*

Before entering upon the discussion of the motion of a rigid body, what is known as the principle of energy will be explained, as it is exceedingly useful, and often gives a partial solution of a problem without any reference to the equations of motion, and in many cases furnishes solutions which are both simple and elegant when compared with those obtained by the use of Cartesian coördinates.

If a single particle of mass m be moving along the axis of x, under the action of a force F in the same direction, we have, as the equation of motion,

$$m\frac{d^2x}{dt^2} = F.$$

And multiplying both sides by $\frac{dx}{dt}$ and integrating, we get

$$\tfrac{1}{2} m (v^2 - V^2) = \int_0^x F dx,$$

where V is the initial value of v or $\frac{dx}{dt}$.

The expression on the left-hand side of the equation is the change in *kinetic energy*, which is equal to the *work done* by the force from o to x.

If, then, we wish to find the length of a simple pendulum which will oscillate in the same time as an extended body, we take

$$l = \frac{h^2 + k^2}{h},$$

which is called the *length of the equivalent simple pendulum*.

Experimentally, l may be found approximately by suspending near the body a simple pendulum made of a small heavy body and a fine string whose length can be adjusted until the times of oscillation of the two are the same.

37. *Centres of Suspension and of Oscillation.*

Let a body be oscillating under the action of gravity about an axis through S perpendicular to the plane of the paper, Fig. 23,

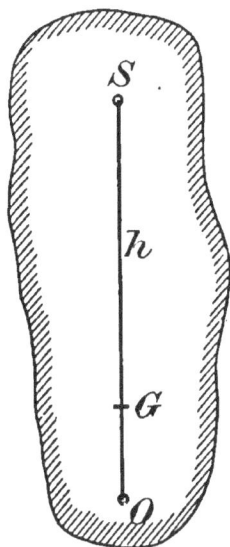

Fig. 23.

and let G be the centre of inertia, and $SO = l = \frac{h^2 + k^2}{h}$, the length of the equivalent simple pendulum. S is called the *centre of suspension*, and O the *centre of oscillation*. Now, if the body be inverted so that it can oscillate about a new axis through O, then the new length l' of the simple equivalent pendulum will be equal to

$$\frac{\left(\frac{k^2}{h}\right)^2 + k^2}{\frac{k^2}{h}} = \frac{k^2 + h^2}{h} = l.$$

Hence, *the centres of suspension and of oscillation are interchangeable.*

38. If the position of the axis of oscillation in a body is changed, the time of oscillation also changes, and it will be found that this time is a maximum when the axis passes through the centre of inertia, and a minimum when $h = k$, and k itself is a minimum. This may be seen either by differentiating the expression for l or by throwing it into the form

$$2\,k + \frac{(h-k)^2}{h}.$$

Illustrative Examples.

1. A cube, edge horizontal and fixed, makes small oscillations about this edge.

If $2\,a$ be the edge, $\qquad l = \frac{4\sqrt{2}}{3}a.$

2. Find the time of a small oscillation of a hemisphere about a horizontal diameter as fixed axis, under gravity.

3. A wire, bent into a circle, oscillates under gravity (1) about a horizontal tangent, (2) about a line perpendicular to this tangent at the point of contact. Compare the times of oscillation.

sum of all the forces acting on the individual particles which compose the body. Thus, for all the particles, we must have

$$\Sigma\left(m\frac{d^2x}{dt^2}\right) = \Sigma(mX) + \Sigma(f_1),$$

$$\Sigma\left(m\frac{d^2y}{dt^2}\right) = \Sigma(mY) + \Sigma(f_2),$$

$$\Sigma\left(m\frac{d^2z}{dt^2}\right) = \Sigma(mZ) + \Sigma(f_3),$$

and D'Alembert's principle states that

$$\Sigma(f_1) = \Sigma(f_2) = \Sigma(f_3) = 0.$$

23. The equations of motion of a rigid body, then, are

$$\left.\begin{aligned}\Sigma\left(m\frac{d^2x}{dt^2}\right) &= \Sigma(mX),\\[2pt]\Sigma\left(m\frac{d^2y}{dt^2}\right) &= \Sigma(mY),\\[2pt]\Sigma\left(m\frac{d^2z}{dt^2}\right) &= \Sigma(mZ).\end{aligned}\right\} \qquad \text{(A)}$$

Each force of the type $m\dfrac{d^2x}{dt^2}$ is termed an *effective force ;* and the above relations are equivalent to saying that the effective forces, if reversed, would be in equilibrium with the external or impressed forces ; they may be looked upon either as equations of motion or as conditions for equilibrium.

24. It is evident, also, from this same principle, that if we take the *sum* of all the moments of the effective forces, these, if reversed, will balance the sum of all the moments of the external forces. Consequently, for any set of rectangular axes, we must also have

$$\Sigma\left[m\left(y\frac{d^2z}{dt^2}-z\frac{d^2y}{dt^2}\right)\right]=L,$$

$$\Sigma\left[m\left(z\frac{d^2x}{dt^2}-x\frac{d^2z}{dt^2}\right)\right]=M,$$

$$\Sigma\left[m\left(x\frac{d^2y}{dt^2}-y\frac{d^2x}{dt^2}\right)\right]=N,$$

(B)

where L, M, N are the couples produced by the external forces.

25. It may be stated here that D'Alembert's principle holds also in the case of a system of bodies moving under their mutual actions and reactions, and applies to the motion of liquids. It is a direct consequence of Newton's Third Law of Motion.

26. *Deductions from D'Alembert's Principle.*

Taking any one of the equations (A), we have

$$\Sigma m\frac{d^2x}{dt^2}=\Sigma mX.$$

But, by definition of the centre of inertia,

$$\Sigma mx=M\bar{x},$$

and $$\therefore\ \Sigma m\frac{d^2x}{dt^2}=M\frac{d^2\bar{x}}{dt^2}.$$

Therefore the above relation becomes

$$M\frac{d^2\bar{x}}{dt^2}=\Sigma mX,$$

and similarly for the other two.

(1) *Hence, the motion of the centre of gravity of a system under the action of any forces is the same as if all the mass were collected at the centre of inertia and all the forces were applied there parallel to their former direction.*

And so the problem of finding the motion of the centre of inertia of a system, however complex, is reduced to finding that of a single particle.

D

plete oscillation given by the relation $t = 2\pi\sqrt{\dfrac{l}{g}}$, where l is the length of the equivalent simple pendulum and equal to $\dfrac{h^2 + k^2}{h}$. If, now, t be observed by means of a clock, and h and k be found, we have the value of g given. This method is one of the most accurate known for finding the intensity of the earth's attraction at different points on its surface. Various forms have been given to these pendulums, from time to time, in order to ensure accuracy of measurement; and the most important of those which have been used for the scientific determination of gravity are described below.

(a) *Borda's Pendulum.*

Borda (1792) constructed his pendulum so as to realize as nearly as possible the simple pendulum. It was made of a sphere of known radius, equal to a. To render it very heavy it was composed of platinum and was suspended by a very fine wire about twelve feet in length. The knife edge which carried the wire and sphere was so arranged by means of a movable screw as to oscillate in the same time as the complete pendulum.

The time was determined by the *method of coincidences*, and g was found from the relation

$$t = 2\pi\sqrt{\dfrac{l + \dfrac{2a^2}{5l}}{g}} \cdot \left(1 + \dfrac{a^2}{16}\right),$$

where l is the length from the knife edge to the centre of the sphere, a the radius of the sphere, and a half the angle through which the pendulum swings at each oscillation to or fro.

(b) *Kater's Pendulum.*

In 1818, Captain Kater determined the value of gravity at London by applying to the pendulum the principle discovered by *Huyghens*, that the centres of suspension and oscillation are reversible. He made a pendulum of a bar of brass about an inch and a half wide and an eighth of an inch in thickness. This bar was pierced in two places, and triangular knife edges of hard steel were inserted so that the distance between them was nearly 39 inches. A large mass in the form of a cylinder was placed near one of the knife edges, being slid on by means of a rectangular opening cut in it. A smaller mass was also attached to the pendulum in such a way as to admit of small motions either way. The pendulum was then swung about the two axes and adjustment of the masses made until the times of small oscillations were the same. This time being noted, and the distance between the knife edges being accurately measured, g was readily calculated. A small difference being generally found in the two times, it can be shown that the length of the seconds pendulum will be found from the expression

$$\frac{(h_1 + h_2)(h_1 - h_2)}{(t_1^2 h_1 - t_2^2 h_2)},$$

where h_1, h_2 are the distances of the centre of inertia from the two knife edges, and t_1, t_2 the corresponding times of oscillation.

(c) *Repsold's Pendulum.*

It was noticed in experimenting with pendulums made like Kater's that the vibration is differently affected by the surrounding air according as the large mass is above or below. This led to the form known as *Repsold's*, in which the two ends are exactly similar externally, but the pendulum (which is cylindrical) is hollow at one end.

The centre of inertia of the figure is equidistant from the knife edges, but the true centre of inertia of the whole mass is at a different point.

40. Many observers have, during the present century, conducted observations at different points on the earth's surface in order to determine not only the length of the seconds pendulum, but also the excentricity of the earth considered as a spheroid.

Helmert in his work on *Geodesy* has collated the results of nearly all the more important expeditions, and the following table gives some of the principal stations with the corresponding lengths of the seconds pendulums there, and the name of the observer. To find g from this table for any place, the relation

$$\log g = 2 \log \pi + \log l$$

PLACE.	LATITUDE.	*l*.	OBSERVER.
Rawak	0° 1′ S.	99.0966	Freycinet
St. Thomas	0 24 N.	99.1134	Sabine
Galapagos	0 32 N.	99.1019	Hall
Para	1 27 S.	99.0948	Foster
Ascension.	7 55 S.	99.1217	———
Sierra Leone. . . .	8 29 N.	99.1104	Sabine
Trinidad	10 38 N.	99.1091	———
Aden	12 46 N.	99.1227	Basevi and Heaviside
Madras	13 4 N.	99.1168	Basevi and Heaviside
St. Helena	15 56 S.	99.1581	———
Jamaica	17 56 N.	99.1497	Sabine
Calcutta	22 33 N.	99.1712	Basevi and Heaviside
Rio Janeiro	22 55 S.	99.1712	———
Valparaiso	33 2 S.	99.2500	Lütke
Montevideo	34 54 S.	99.2641	Foster
Lipari	38 28 N.	99.3097	Biot
Hoboken, N.J. . . .	40 44 N.	99.3191	———
Tiflis	41 41 N.	99.3190	———
Toulon	43 7 N.	99.3402	Duperrey
Bordeaux	44 50 N.	99.3470	Biot
Padua	45 24 N.	99.3623	Biot
Paris	48 50 N.	99.3858	———
Shanklin Farm (Isle of Wight)	50 37 N.	99.4042	Kater
Kew	51 28 N.	99.4169	———
Greenwich	51 28 N.	99.4143	———
London	51 31 N.	99.4140	———
Berlin	52 30 N.	99.4235	———
Staten Island . . .	54 46 S.	99.4501	Foster
Cape Horn	55 51 S.	99.4565	Foster
Leith	55 58 N.	99.4550	———
Sitka	57 3 N.	99.4621	Lütke
Pulkowa	59 46 N.	99.4854	Sawitsch
Petersburg	59 56 N.	99.4876	———
Unst	60 45 N.	99.4959	———

may be used, where l is the length of the seconds pendulum in *centimetres*. See also *Geodesy, by Colonel A. R. Clarke*, Chap. XIV.

The places are arranged geographically in order of their latitudes, and show thereby the gradual increase in the length of the seconds pendulum as we go from the equator to the pole.

Those places, in the preceding table, for which the lengths of the seconds pendulum have been calculated from a number of observations made by different observers, are indicated by a dash.

41. During the past few years several observers have made observations on the value of g at different points in North America. Professor Mendenhall, of the U. S. Coast Survey, during the summer of 1891, visited a number of places on the Pacific coast between San Francisco and the coast of Alaska, and in his report of the expedition gives a table of the values determined, with the places and corresponding latitudes. He made use of a half-seconds pendulum enclosed in an air-tight chamber which could be exhausted with an air pump. A special method was used for noting the coincidences (see *U. S. Coast and Geodetic Survey. Report for* 1891, Part 2).

Defforges, one of the greatest living authorities on methods of gravity determination, crossed from Washington to San Francisco during the summer of 1893 and made a number of observations which are given in the following table. The value of g alone is given.

Washington	980.169
Montreal	980.747
Chicago	980.375
Denver	980.983
Salt Lake City	980.050
Mt. Hamilton	979.916
San Francisco	980.037

These are all reduced to sea level.

42. *Experimental Determination of a Moment of Inertia.*

In many cases of small oscillations under gravity, where it
is difficult to calculate the moment of inertia of a body from its
elements, the time of oscillation is observed ; and, the moment
of inertia being increased by the addition of a mass of definite
figure, the time of oscillation is again noted.

The required moment of inertia may then be calculated.

This method is particularly useful in the case of magnetic
oscillations about a vertical axis.

43. *Pressure on the Fixed Axis. Forces and Body Symmetrical.*

If a body be moving about an axis, and it is symmetrical with
respect to a plane passing through the centre of inertia and
perpendicular to the axis, and at the same time the forces acting
on the body are also symmetrical with respect to this plane,
then we may suppose that the pressures on the axis are reducible to a single one which will lie in the plane of symmetry and
will cut the axis of rotation. To determine, in such case, the
direction and magnitude of the resultant pressure, we proceed
in the following way.

Let the body, Fig. 25, surround the point O and let it be
symmetrical with respect to the plane of the paper which contains C, the centre of inertia : the axis of rotation being perpendicular to the plane of the paper, and passing through O. Let
the forces acting on the body also be symmetrical with reference
to this plane. And let the body, moving about the axis through
O, be situated at any time t as represented, θ being the angle
which the line OC fixed in the body and moving with it makes
with the line OA fixed in space. Then the resultant pressure
on the axis will be in the plane of the paper, and its direction
will pass through O. Let its components measured along two
rectangular axes OX, OY in the body, be P and Q. Let $CO=h$.

Then, X, Y, being the accelerations on unit mass in the

directions OX, OY, we have, by D'Alembert's principle, the relations

$$\Sigma m \frac{d^2x}{dt^2} = \Sigma mX + P,$$

$$\Sigma m \frac{d^2y}{dt^2} = \Sigma m Y + Q.$$

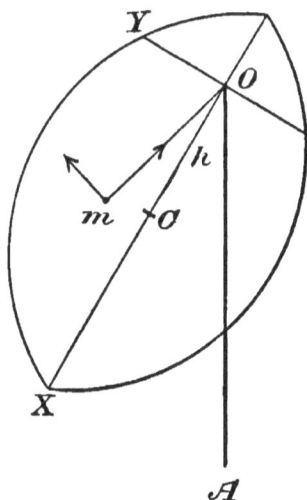

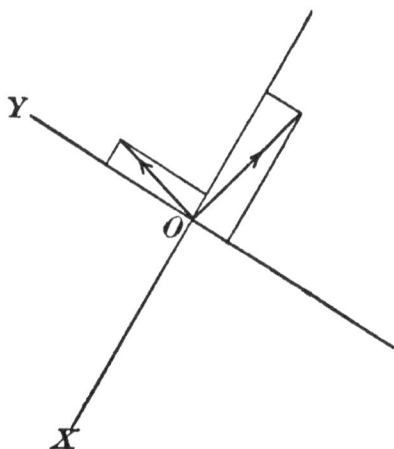

Fig. 25.   Fig. 26.

Now, if ω be the angular velocity, any particle such as m will be acted on by the forces $m\omega^2 r$, $m\dot{\omega} r$, as is indicated in the figure ; and these forces resolved along OX, OY, as shown in Fig. 26, would give

$$m\frac{d^2x}{dt^2} = .- m\omega^2 x - m\dot{\omega} y,$$

$$m\frac{d^2y}{dt^2} = - m\omega^2 y + m\dot{\omega} x.$$

The values of $\frac{d^2x}{dt^2}$, $\frac{d^2y}{dt^2}$ may also be obtained by direct differentiation from $x = r \cos \theta$, $y = r \sin \theta$. Thus,

$$\frac{dx}{dt} = -r \sin \theta \frac{d\theta}{dt} = -y\omega, \qquad \frac{dy}{dt} = r \cos \theta \frac{d\theta}{dt} = x\omega.$$

$$\therefore \frac{d^2x}{dt^2} = -y\dot{\omega} - \omega^2x, \qquad \frac{d^2y}{dt^2} = x\dot{\omega} - \omega^2y.$$

Hence, our relations for determining the pressures become

$$P + \Sigma mX + \Sigma m(\omega^2x + \dot{\omega}y) = 0,$$

$$Q + \Sigma mY + \Sigma m(\omega^2y - \dot{\omega}x) = 0.$$

$$\therefore P = -\Sigma mX - \Sigma m(\omega^2x + \dot{\omega}y),$$

$$Q = -\Sigma mY - \Sigma m(\omega^2y - \dot{\omega}x).$$

But, by definition of the centre of inertia,

$$\Sigma m\omega^2x = \omega^2\Sigma mx = Mh\omega^2, \quad \Sigma m\dot{\omega}y = \dot{\omega}\Sigma my = 0,$$

$$\Sigma m\omega^2y = \omega^2\Sigma my = 0, \quad \Sigma m\dot{\omega}x = \dot{\omega}\Sigma mx = Mh\dot{\omega}.$$

$$\therefore P = -\Sigma mX - Mh\omega^2,$$

$$Q = -\Sigma mY + Mh\dot{\omega},$$

which equations determine the pressures P, Q, and therefore the direction and magnitude of the resultant pressure when we know ω, which is found from the relation already given,

$$\dot{\omega} = \frac{d^2\theta}{dt^2} = \frac{N}{\Sigma mr^2},$$

where N is the moment of the external forces about the rotation axis, and Σmr^2 is the moment of inertia about the same axis. This, on integration, gives ω, and on substituting its value in the preceding expression, P and Q are found.

44. *Heavy Symmetrical Body. Pressure on the Axis.*

In the particular case of a *heavy* body which is symmetrical about a plane through its centre of inertia perpendicular to the rotation axis, which is horizontal, the external forces are only those of gravity, and we have, Fig. 27, the pressures given by the relations

$$P = -Mg \cos \theta - Mh\omega^2,$$

$$Q = Mg \sin \theta + Mh\dot\omega,$$

and if we suppose P estimated in the opposite direction, the complete solution of the motion is obtained from

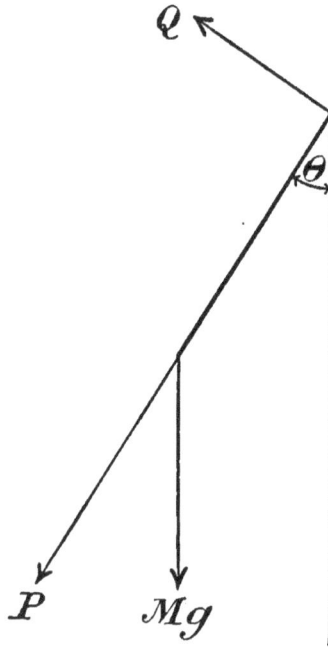

Fig. 27.

$$\left.\begin{array}{l}
\dfrac{d\omega}{dt} = \dfrac{d^2\theta}{dt^2} = -\dfrac{gh \sin \theta}{h^2 + k^2}, \\[2mm]
P = Mg \cos \theta + Mh\omega^2, \\[1mm]
Q = Mg \sin \theta + Mh\dot\omega,
\end{array}\right\}$$

θ being measured always upwards from the vertical and k being the radius of gyration about the centre of inertia.

Illustrative Examples.

1. A rod, movable about one end, falls in a vertical plane, starting from a horizontal position. Find the pressure on the end in any position.

Figure 28 shows the motion; when the rod makes an angle θ with the vertical line OA, we have

$$\frac{d\omega}{dt}=\frac{d^2\theta}{dt^2}=-\frac{ga\sin\theta}{\frac{4\,a^2}{3}}=-\frac{3\,g}{4\,a}\sin\theta.$$

$$\therefore\ 2\,\omega\frac{d\omega}{dt}=-\frac{3\,g}{2\,a}\sin\theta\frac{d\theta}{dt}.$$

$$\therefore\ \int_0^\omega 2\,\omega d\omega=-\int_{\frac{\pi}{2}}^\theta\frac{3\,g}{2\,a}\sin\theta d\theta.$$

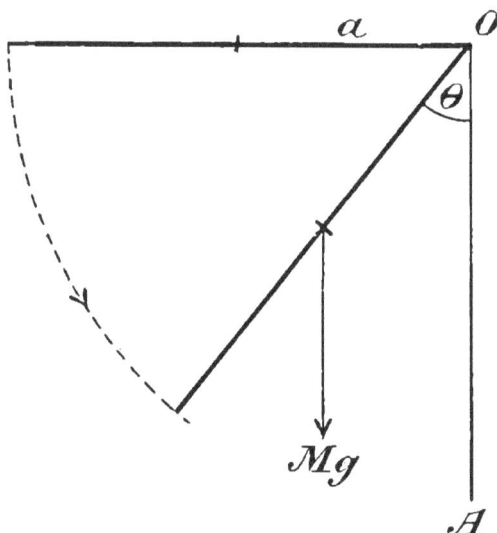

Fig. 28.

$$\therefore\ \omega^2=\frac{3\,g}{2\,a}\cos\theta,$$

and
$$P=Mg\cos\theta+Ma\omega^2=\tfrac{5}{2}\,Mg\cos\theta,$$
$$Q=Mg\sin\theta+Ma\dot\omega=\tfrac{1}{4}\,Mg\sin\theta.$$

When the rod is in the lowest position, $\theta = 0$, and $P = \frac{5}{2} Mg$, $Q = 0$.

2. Rod, movable about one end, falling from the position of unstable equilibrium.

As in the preceding problem, we have (Fig. 29)

$$\frac{d\omega}{dt} = -\frac{3g}{4a} \sin \theta,$$

$$\int_0^\omega 2\,\omega d\omega = -\int_\pi^\theta \frac{3g}{2a} \sin \theta d\theta,$$

$$\omega^2 = \frac{3g}{2a}(1 + \cos \theta),$$

and $\qquad P = Mg \cos \theta + Ma\omega^2 = \frac{1}{2} Mg(3 + 5 \cos \theta),$

$$Q = Mg \sin \theta + Ma\dot{\omega} = \frac{1}{4} Mg \sin \theta.$$

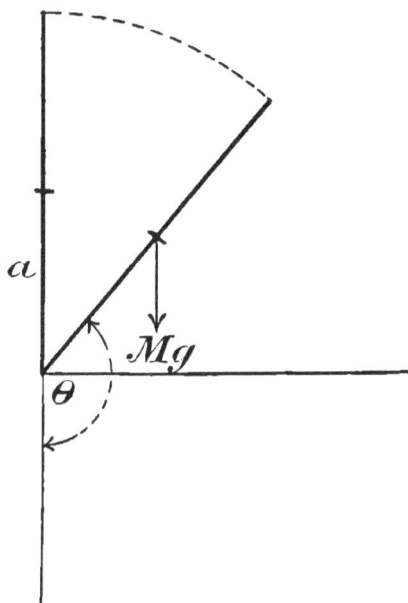

Fig. 29.

In the lowest position, $\theta = 0$, $Q = 0$, $P = 4 Mg$, which shows that if the rod can just make complete revolutions, the pressure on the axis in the lowest position is in the direction of the rod, and equal to four times its weight.

Maximum and Minimum Values of P and Q.

P is a maximum when $\theta = 0$ or π, and its values then are $4\,Mg$ and $-Mg$; it is a minimum when $\cos\theta = -\frac{3}{5}$. Q is a maximum when $\theta = \frac{\pi}{2}$, and its value then is $\frac{1}{4}\,Mg$; it is a minimum when $\theta = 0$ or π, and its value then is 0.

Resultant Pressure at Any Time.

This may be found by taking $R^2 = P^2 + Q^2$, and substituting the general values of P and Q in terms of θ. The maximum and minimum values of the total pressure may be obtained by differentiating in the usual way. The angle which the resultant pressure makes with the rod will be determined from the relation $\tan\psi = \dfrac{Q}{P} = \dfrac{\sin\theta}{6 + 10\cos\theta}$.

3. Cube, edge horizontal, performing complete revolutions under gravity.

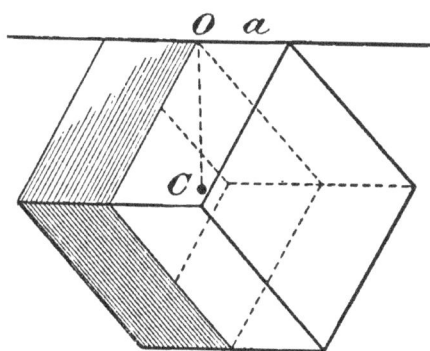 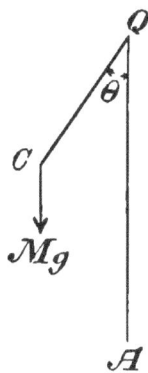

Fig. 30.   Fig. 31.

Figs. 30 and 31 show the motion. Since the body and forces are symmetrical about the central plane perpendicular to the axis of rotation, the pressures on the axis, as the cube swings around, are reducible to a single pressure lying in this central plane and cutting the axis. Taking, then, the auxiliary figure, we need only consider the motion of OC, which in any position makes an angle θ with the vertical line OA.

The angular velocity at any instant is given by

$$\frac{d\omega}{dt} = \frac{d^2\theta}{dt^2} = -\frac{3g}{4\sqrt{2}\,a}\sin\theta,$$

the edge of the cube being of length $2\,a$.

Supposing the cube to start initially with OC vertically upwards, and to swing completely around,

$$\therefore \int_0^\omega 2\,\omega\,d\omega = -\int_\pi^\theta \frac{3g}{2\sqrt{2}\,a}\sin\theta\,d\theta,$$

$$\therefore \omega^2 = \frac{3g}{2\sqrt{2}\,a}(1+\cos\theta),$$

and

$$P = Mg\cos\theta + Ma\sqrt{2}\left(\frac{3g}{2\sqrt{2}\,a}\overline{1+\cos\theta}\right),$$

$$Q = Mg\sin\theta + Ma\sqrt{2}\left(-\frac{3g}{4\sqrt{2}\,a}\sin\theta\right).$$

$$\therefore P = \frac{Mg}{2}(3+5\cos\theta),$$

$$Q = \frac{Mg}{4}\sin\theta.$$

The maximum and minimum values of P and Q can easily be found, as in the previous case of the rod.

To find the *total pressure*, we have

$$R^2 = P^2 + Q^2 = \left(\frac{Mg}{2}\right)^2(3+5\cos\theta)^2 + \left(\frac{Mg}{4}\right)^2\sin^2\theta,$$

and the maximum and minimum values of R can be found by the process of differentiation with respect to θ. It will be

F

found that R is maximum when $\theta = 0$ or $n\pi$, and its values then are $4\,Mg$ and $-Mg$.

R is a minimum when $\cos\theta = -\frac{20}{33}$, and its value is $\dfrac{Mg}{4}\sqrt{\dfrac{7}{11}}$.

4. A hemisphere revolves about an axis which coincides with a diameter of its base, and which is inclined at an angle α to the vertical. If it swings completely around, the total pressure on its axis, when in the lowest position, is

$$\frac{W}{64}\{(109\sin\alpha)^2 + (64\cos\alpha)^2\}^{\frac{1}{2}}.$$

5. A right circular cone whose height is equal to the radius of its base swings about a horizontal axis through its vertex. If the axis of the cone starts from a horizontal position, find the angular velocity and the pressure on the rotation axis when in the lowest position.

6. A uniform heavy rod oscillates about one end in a vertical plane, under gravity, coming to rest in a horizontal position. If ψ be the angle between the rod and the line of the resultant pressure, and θ the angle of inclination of the rod to the horizon at the same time, then $\tan\psi\,\tan\theta = \frac{1}{10}$.

7. A homogeneous solid spheroid, the equation of whose bounding surface is

$$\frac{x^2}{a^2} + \frac{y^2+z^2}{b^2} = 1,$$

is suspended from an axis passing through one of the foci. Prove that the centre of oscillation lies on the surface

$$\{a^2x^2 + b^2(x^2+y^2+z^2)\}^2 = 25\,a^2(a^2-b^2)(x^2+y^2+z^2)^2.$$

8. A uniform wire is bent into the form of an isosceles triangle, and revolves about an axis through its vertex perpendicu-

lar to its plane. Prove that the centre of oscillation will be at the least possible distance when the triangle is right angled.

9. A uniform heavy rod revolves uniformly about one end in such a manner as to describe a cone of revolution. Find the pressure on the fixed point, and show that if θ, ψ be the angles which the vertical makes with the rod and the resultant pressure, $4 \tan \psi = 3 \tan \theta$.

10. A rough uniform board is placed on a horizontal table with two-thirds of its length projecting over the table, the board being initially in contact with the table, and perpendicular to the edge. Show that it will begin to slide off when it has turned through an angle $\tan^{-1} \dfrac{\mu}{2}$, μ being the coefficient of friction.

45. *General Case.*

If the forces and body are not symmetrical, then we take the general equations already found; and supposing the pressures to be equivalent to two at two points on the axis whose components are P, Q, R; P', Q', R'; we get, for the determination of these pressures, the relations

$$\Sigma m \frac{d^2x}{dt^2} = \Sigma m X + P + P',$$

$$\Sigma m \frac{d^2y}{dt^2} = \Sigma m Y + Q + Q',$$

$$\Sigma m \frac{d^2z}{dt^2} = \Sigma m Z + R + R',$$

$$\Sigma m \left(y \frac{d^2z}{dt} - z \frac{d^2y}{dt^2} \right) = L + C_1 + C_2,$$

$$\Sigma m \left(z \frac{d^2x}{dt^2} - x \frac{d^2z}{dt^2} \right) = M + C_1' + C_2',$$

$$\Sigma m \left(x \frac{d^2y}{dt^2} - y \frac{d^2x}{dt^2} \right) = N = \Sigma m r^2 \cdot \frac{d\omega}{dt}.$$

This last relation gives at once the value of $\dot{\omega}$ and, by integration, of ω. L, M, N are the couples produced by the external forces; and C_1, C_2, C_1', C_2', the couples produced by the pressures, which can be expressed in terms of these pressures, and the distances from the origin at which they are supposed to act.

The process of solving any particular problem will be to

(1) Find $\dot{\omega}$ and ω.

(2) Express the quantities $\dfrac{d^2x}{dt^2}$, etc., in terms of $\dot{\omega}$, ω, and known expressions.

(3) Thence find the pressures.

The effective forces can be expressed in terms of the radial and transversal forces either by resolution or by direct differentiation, and it will be found that

$$\frac{d^2x}{dt^2} = -\omega^2 x - \dot{\omega}y,$$

$$\frac{d^2y}{dt^2} = -\omega^2 y + \omega x.$$

Thus, the previous relations will become

$$\Sigma mX + P + P' = \Sigma m(-\omega^2 x - \dot{\omega}y) = -\omega^2 M\bar{x} - \dot{\omega}M\bar{y},$$

$$\Sigma mY + Q + Q' = \Sigma m(-\omega^2 y + \dot{\omega}x) = -\omega^2 M\bar{y} + \dot{\omega}M\bar{x},$$

$$\Sigma mZ + R + R' = 0.$$

where $\bar{x}$, $\bar{y}$ are the coördinates of the centre of inertia.

Also, since $\dfrac{d^2z}{dt^2} = 0$, we have

$$L + C_1 + C_2 = \Sigma m\left(y\frac{d^2z}{dt^2} - z\frac{d^2y}{dt^2}\right) = \omega^2 \Sigma myz - \dot{\omega}\Sigma mxz,$$

$$M + C_1' + C_2' = \Sigma m\left(z\frac{d^2x}{dt^2} - x\frac{d^2z}{dt^2}\right) = -\omega^2 \Sigma mxz - \dot{\omega}\Sigma myz,$$

$$N = \Sigma m\left(x\frac{d^2y}{dt^2} - y\frac{d^2x}{dt^2}\right) = \Sigma mr^2 \cdot \frac{d^2\theta}{dt^2}.$$

46. It will be seen that the last expressions are much simplified if we make a proper choice of axes. The first thing to be done, then, is to choose the origin on the rotation axis so thát it is a principal axis at that point. Then $\Sigma mxz = 0$, $\Sigma myz = 0$.

Thus, for example, if we suppose a triangle to be rotating under gravity about one side which is horizontal, the equations of motion will be much simplified if we choose as origin that point in the side at which it is a principal axis ; see Ex. 3, p. 28. Then, supposing the pressures to be equivalent to two acting at the ends of the side, the solution is very simple, as the angular velocity at any time is found from the relation

$$\frac{d\omega}{dt} = -g \frac{\frac{p}{3}\sin\theta}{\frac{p^2}{6}} = -\frac{2g\sin\theta}{p},$$

where p is the perpendicular on the rotation axis from the opposite angle ; and the pressures can then be immediately written down in terms of ω, $\dot{\omega}$ and the coördinate $\bar{x}$ of the centre of inertia, $\bar{y}$ being 0, since the body is a lamina.

CHAPTER V.

47. *General Case.*

An impulse being defined, as already explained, to be a force which produces a sudden change of velocity, and which only acts for an indefinitely short time, we can obtain the general impulsive equations of motion of any body capable of motion about a fixed axis by considering the relations found in Art. 45. In those relations, by the substitution of changes of velocity for accelerations, we get

$$\Sigma X + P + P' = \Sigma m \left\{ \left(\frac{dx}{dt}\right)' - \left(\frac{dx}{dt}\right) \right\},$$

$$\Sigma Y + Q + Q' = \Sigma m \left\{ \left(\frac{dy}{dt}\right)' - \left(\frac{dy}{dt}\right) \right\},$$

$$\Sigma Z + R + R' = 0,$$

where X, Y, Z are the impulsive actions on individual particles due to external impulsive forces; P, Q, R, P', Q', R' are impulsive pressures on the fixed axis; and where the velocity $\left(\frac{dx}{dt}\right)$ of any particle before the impulsive action takes place, is changed suddenly to $\left(\frac{dx}{dt}\right)'$.

And, since $\frac{dz}{dt} = 0$, and the angular velocity ω is suddenly changed to ω', we have for the impulsive couples,

$$L + C_1 + C_2 = -\Sigma mz \left\{ \left(\frac{dy}{dt}\right)' - \left(\frac{dy}{dt}\right) \right\},$$

$$M + C_1' + C_2' = \Sigma mz \left\{ \left(\frac{dx}{dt}\right)' - \left(\frac{dx}{dt}\right) \right\},$$

$$N = (\omega' - \omega) \Sigma mr^2.$$

Now we have $\frac{dx}{dt}$ at any time equal to $-\omega y$, and $\frac{dy}{dt} = \omega x$; and, substituting these values in the preceding equations, we have the complete solution of the problem given by

$$\Sigma X + P + P' = -\Sigma m(\omega' - \omega)y = -(\omega' - \omega)M\bar{y},$$

$$\Sigma Y + Q + Q' = \Sigma m(\omega' - \omega)x = (\omega' - \omega)M\bar{x},$$

$$\Sigma Z + R + R' = 0,$$

$$L + C_1 + C_2 = -\Sigma mz(\omega' - \omega)x = -(\omega' - \omega)\Sigma mxz,$$

$$M + C_1' + C_2' = \Sigma mz(\omega' - \omega)y = (\omega' - \omega)\Sigma myz,$$

$$N = (\omega' - \omega) \cdot \Sigma mr^2.$$

48. If the body starts from rest, then $\omega = 0$, and the sudden angular velocity generated by an impulse which tends to turn a body about a fixed rotation axis is obtained from the relation

$$\omega' = \frac{N}{\Sigma mr^2},$$

where N is the moment of the impulse about the axis, and Σmr^2 is the moment of inertia. As before, the problem is simplified by choosing the origin at a point where the rotation axis is a principal axis.

49. *Centre of Percussion.*

In the general equations just found, let us suppose that the impulsive actions are those caused by a blow Q represented by components X, Y, Z; and that the blow is struck at some point on the surface of a body, capable of motion about a fixed axis, which either passes through it or to which it is rigidly connected. What is the condition that there shall be no impulsive pressure on the axis? Or, in other words, is it possible to strike the body at a certain point in such a way as to produce

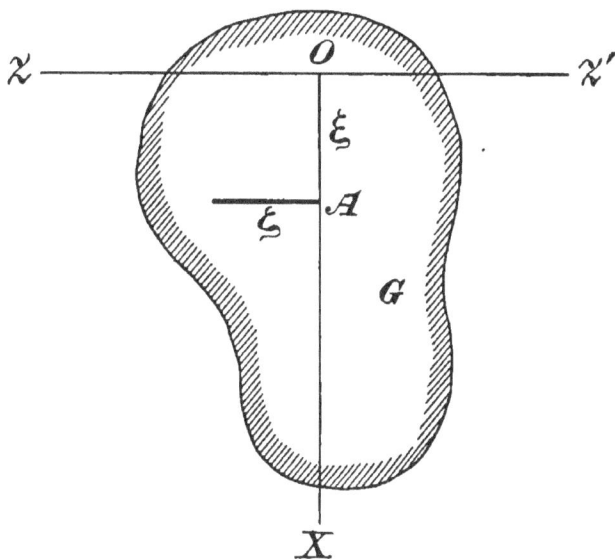

Fig. 32.

no strain upon the axis about which it is free to rotate? Let the body (Fig. 32) surround O; let ZZ' be the axis of rotation, and let the plane of zx, which is the plane of the paper, contain G, the centre of inertia of the body. Suppose that the blow Q is applied at the point whose coördinates are ξ, η, ζ (the coördinate η not being shown, being drawn upwards perpendicular to the plane of the paper). If there be no resulting pressure when the body is struck, the general relations become :

$$X = 0,$$

$$Y = (\omega' - \omega) M\bar{x},$$

$$Z = 0,$$

$$L = \eta Z - \zeta Y = - (\omega' - \omega) \Sigma mxz,$$

$$M = \zeta X - \xi Z = - (\omega' - \omega) \Sigma myz,$$

$$N = \xi Y - \eta X = (\omega' - \omega) \Sigma mr^2 = (\omega' - \omega) Mk^2,$$

where k is the radius of gyration about the axis.

From these it will be seen that, since $X = 0$, $Z = 0$, we have also $\Sigma myz = 0$. And also,

$$\zeta Y = (\omega' - \omega) \Sigma mxz,$$

$$Y = (\omega' - \omega) M\bar{x}.$$

$$\therefore \quad \zeta = \frac{\Sigma mxz}{M\bar{x}} = \frac{\Sigma mxz}{\Sigma mx}.$$

And ξ is given by the last relation,

$$\xi = \frac{(\omega' - \omega) Mk^2}{Y} = \frac{(\omega' - \omega) Mk^2}{(\omega' - \omega) M\bar{x}} = \frac{k^2}{\bar{x}}.$$

The above conditions holding, and there being no pressure on the axis, the line of the blow is called a *Line of Percussion*, and any point in this line is termed a *Centre of Percussion*.

50. By an inspection of the foregoing relations, we have,

1. $X = 0$, $Z = 0$; and therefore one condition, that there may be no strain upon the axis, is that the line of the blow must be perpendicular to the plane containing the rotation axis and the centre of inertia.

2. $\Sigma myz = 0$, and $\Sigma mxz = \zeta \cdot \Sigma mx$.

Now, since O may be chosen anywhere on the axis, let it be so chosen that $\zeta = 0$. Then for that origin so chosen Σmyz would be zero, and Σmxz also zero.

Therefore, an essential condition, to be first satisfied for a line of percussion, is that the axis of rotation must be a principal axis at some point of its length.

3. $\xi = \dfrac{k^2}{x}$, which shows that when a centre of percussion does exist, its distance from the axis is the same as that of the centre of oscillation.

If $\zeta = 0$ and $\bar{z} = 0$, then the line of percussion passes through the centre of oscillation, which may be stated in the following way :

If the fixed axis be parallel to a principal axis at the centre of inertia, the line of action of the blow will pass through the centre of oscillation.

Illustrative Examples.

1. A uniform rod, fixed at one end and capable of motion in a vertical plane, is hanging freely under the action of gravity, and being struck perpendicular to its length, rises into the position of unstable equilibrium. Find the magnitude of the blow that there may be no strain at the fixed point.

In order that there may be no strain on the axis, it must be struck at the centre of percussion, which point will be at a distance $\dfrac{4a}{3}$ from the fixed end, if the length of the rod be $2a$. Then, if ω be the angular velocity produced by the impulse, we have from the equation of moments,

$$B \cdot \frac{4a}{3} = \Sigma mr^2 \cdot \omega.$$

$$\therefore B = Ma\omega.$$

Also, $\dfrac{d\omega}{dt} = -\dfrac{3g}{4a}\sin\theta$

is the equation of motion of the rod as it rises upwards, being acted upon by gravity, and starting with an angular velocity ω.

$$\therefore \int_\omega^0 2\,\omega d\omega = -\frac{3g}{2a}\int_0^\pi \sin\theta d\theta,$$

$$\omega^2 = \frac{3g}{a}.$$

$$\therefore B = Ma\omega = M\sqrt{3ga}.$$

From this it may be seen that generally when a body is struck at the centre of percussion, the value of the impulse is measured by the product of the mass and the velocity of the centre of inertia.

2. A circular plate free to move about a horizontal tangent is stuck at its centre of percussion and rises into a horizontal position. Find the blow.

As before, $B = Ma\omega$, a being the radius,

and $\dfrac{d\omega}{dt} = -\dfrac{4g}{5a}\sin\theta$ gives ω.

$$\therefore B = M\sqrt{\frac{8ga}{5}}.$$

3. A sector of a circle, whose radius is a and angle α, is capable of turning about an axis in its plane which is perpendicular to one of its bounding radii. Find the coördinates of the centre of percussion.

Fig. 33 shows the position of the centre of percussion C, whose coördinates are

$$\zeta = \frac{\Sigma mxz}{\Sigma mx},$$

$$\xi = \frac{\Sigma mx^2}{\Sigma mx}.$$

On transforming to polar coördinates it will be found that

$$\zeta = \tfrac{3}{8} a \sin \alpha,$$

$$\xi = \tfrac{3}{8} a \left(\frac{\alpha}{\sin} + \cos \alpha \right).$$

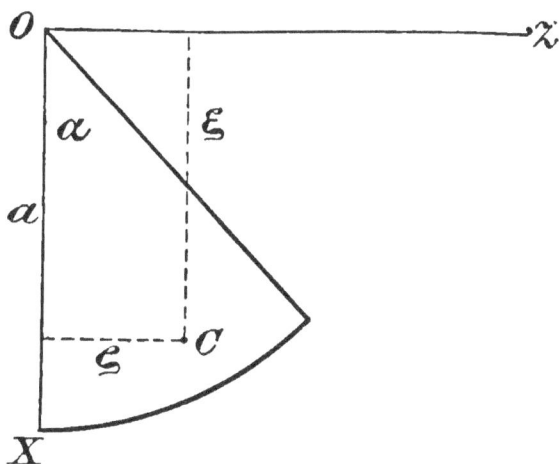

Fig. 33.

4. To find the centre of percussion of a triangular plate capable of rotation about a side.

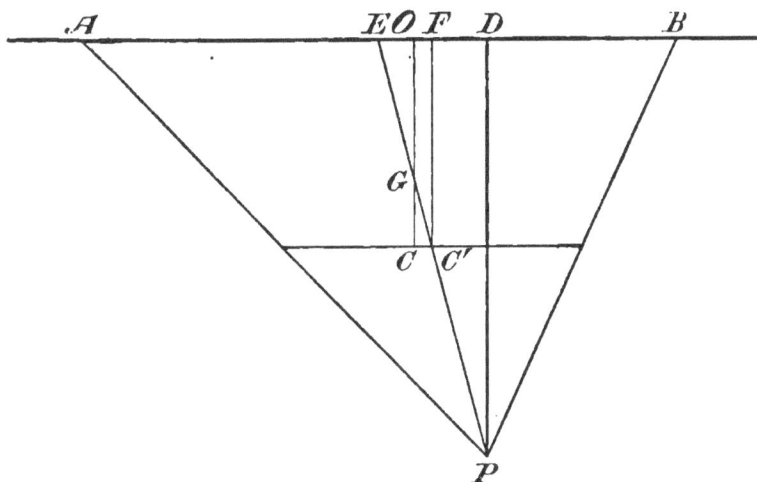

Fig. 34.

Fig. 34 shows the position of the centre of percussion. AB is the rotation axis, PD perpendicular to AB, E the middle point of AB, F the middle point of DE. Then AB is a principal axis at the point F, and G being the centre of inertia of the plate, and $PD=p$,

C is the centre of oscillation,

C' is the centre of percussion,

and
$$OC=\frac{k^2}{h}=\frac{\frac{p^2}{6}}{\frac{p}{3}}=\frac{p}{2}.$$

When the triangle is isosceles, C and C' coincide.

5. $ABCD$ is a quadrilateral (Fig. 35), AB being parallel to CD. Show that, if $AB^2=3CD^2$, the point P is a centre of percussion for the rotation axis AB. (Wolstenholme.)

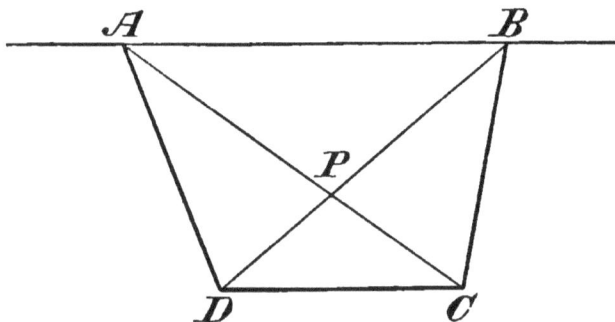

Fig. 35.

6. A uniform beam capable of motion about one end is in equilibrium. Find at what point a blow must be applied perpendicular to the beam in order that the impulsive action on the fixed end may be one-third of the blow.

51. *Initial Motions.   Changes of Constraint.*

If a body, moving about a fixed axis with known angular
velocity, is suddenly freed from its constraint and a new axis
fixed in it, or if a body at rest is disturbed so that there is a
sudden impulsive change of pressure, we can determine the
new angular velocities and changes of pressure by reference to
the impulsive equations of motion already found.   Sometimes,
however, solutions which are more instructive may be obtained
by considering elementary principles; and the following examples are given to illustrate the methods to be employed in
various cases.

Illustrative Examples.

1. A uniform board is placed on two props; if one be suddenly removed, find the sudden change in pressure at the other.

Fig. 36 illustrates the problem.   The board is of length $2a$,
and rests on the props A and B, which are fixed in position in

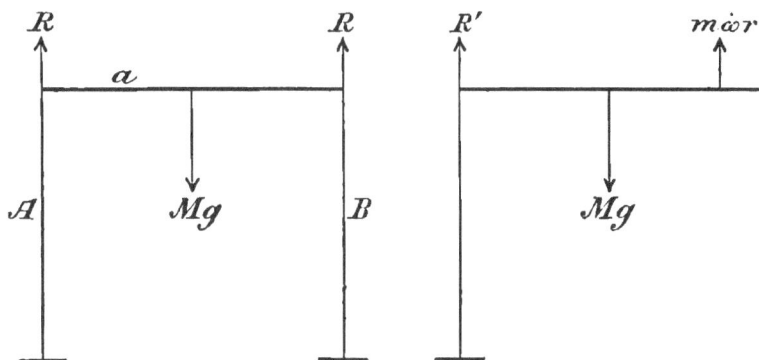

Fig. 36.

the first figure, so that $R = \frac{1}{2} Mg$.   If B be now removed, the
board begins to turn about the upper end of A under the action
of gravity, and to each element of the board an acceleration $\dot{\omega}r$
is given suddenly; so that if we communicated to each element
m an acceleration $\dot{\omega}r$ in the opposite direction (upwards), we

would have, by the application of D'Alembert's principle, R', $\Sigma(m\dot\omega r)$, and Mg in equilibrium with one another, as indicated in the second figure.

$$\therefore\ R' + \Sigma(m\dot\omega r) - Mg = 0.$$

Also, taking moments about O, just when the prop is removed, we have

$$\Sigma(m\dot\omega r)\cdot r = Mg\cdot a.$$

$$\therefore\ \dot\omega = \frac{3}{4}\frac{g}{a}.$$

$$\therefore\ R' = Mg - Ma\dot\omega = \tfrac{1}{4}\,Mg.$$

2. The extremities of a heavy rod are attached by cords of equal length to a horizontal beam, the cords making an angle of 30° with the beam. If one of the cords be cut, show that the initial tension of the other is two-sevenths of the weight of the rod.

3. A uniform rod is suspended in a horizontal position by means of two strings which are attached to the ends of the rod. If one of these strings be suddenly cut, find the sudden change in tension of the other string.

4. Two strings of equal length have each an extremity tied to a weight C, and their other extremities tied to two points A, B in the same horizontal line. If one be cut, the tension of the other is instantaneously altered in the ratio $1 : 2\cos^2\dfrac{ACB}{2}$.

5. A particle is suspended by three equal strings of length a from three points forming an equilateral triangle of side $2b$ in a horizontal plane. If one string be cut, the tension of each of the others is instantaneously changed in the ratio $\dfrac{3a^2 - 4b^2}{2(a^2 - b^2)}$.

6. A rod of length $2a$ falls from a vertical position, being capable of motion about one end in a vertical plane, and when in a horizontal position, strikes a fixed obstacle at a given distance from the end. Find the magnitude of the impulse, and the pressure on the fixed end.

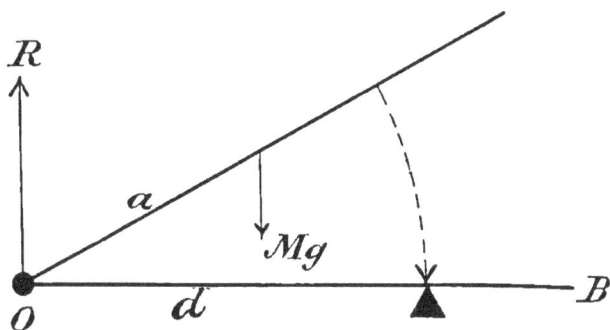

Fig. 37.

Let the rod (Fig. 37) drop from the vertical position and strike an obstacle when in the position OB with a blow Q. Let R be the impulse on the fixed end O, and then we have, taking moments about O,

$$Qd = \Sigma(mr\omega) \cdot r = Mk^2 \cdot \omega.$$

$$\therefore Q = \frac{M \cdot \dfrac{4a^2}{3} \omega}{d},$$

and since the rod falls from the vertical position, its angular velocity when in the horizontal position is found in the usual way to be given by

$$\omega^2 = \frac{3g}{2a}.$$

$$\therefore Q = \frac{2Ma}{d}\sqrt{\frac{2ag}{3}}.$$

· The impulsive pressure on the fixed end is obtained from the relation

$$Q = R + \Sigma(mr\omega) = R + Ma\omega.$$

$$\therefore R = Q - Ma\omega = Ma\omega \left\{ \frac{4a}{3d} - 1 \right\}.$$

$$\therefore R = Ma \sqrt{\frac{3g}{2a}} \left\{ \frac{4a}{3d} - 1 \right\}.$$

These two values of Q and R will change as d changes; Q will be a maximum when $d = \frac{4a}{3}$, and R will be positive, zero, or negative, according as

$$4a > = < 3d,$$

or as $$d < = > \frac{4a}{3}.$$

Hence, if the obstacle is beyond the centre of percussion, the impulsive strain at O is downwards. If at the centre of percussion there is no impulsive action on the axis, and when $d < \frac{4a}{3}$, the impulse at O is upwards.

These results can easily be verified by experiment. An iron bar movable about an axis in which it is very loosely held, if dropped so that it strikes an obstacle in a horizontal position, will throw its fixed end downwards or upwards according as the obstacle is beyond or nearer than the centre of percussion; and if the bar falls so that it strikes the obstacle just at the centre of percussion, then there is no jar on the fixed end, no matter how loosely it may be held. The experiment may be modified in many ways, and a familiar illustration of there being a centre of percussion is afforded by the use of the cricket bat or base-ball club with which a ball is struck. If the ball be struck by a portion of the bat out near the end, the fingers tingle from the impulsive reaction outwards; if it be struck nearer than the centre of percussion, the impulsive reaction is inwards against the palm of the hand; when the ball is struck properly, there is no impulsive reaction on the hand, and the energy is all communicated to the ball.

G

7. A rod is moving with uniform angular velocity about one end fixed; suddenly this end is freed and the other end fixed. Find the new angular velocity.

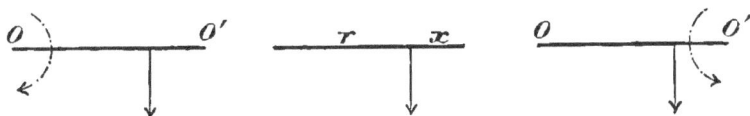

Fig. 38

Fig. 38 indicates the solution. In the first figure each particle has a linear velocity ωr in the direction indicated, on account of the angular velocity ω. In the second figure both ends are free, and the velocities remain as before. In the third figure O' is instantaneously fixed, which does not affect the velocities of the other elements of the rod, by the definition of an impulse. And hence ω', the new angular velocity about O', will be as shown in the figure in direction, and its magnitude will be found by using the formula for moment of momentum. Thus

$$\Sigma(m\omega r)x = \Sigma(m\omega'x)x.$$

And if $x + r = a$, and ρ is the density,

$$\therefore \ \omega \int_0^a \rho(a-x)x\,dx = M \cdot \frac{a^2}{3}\omega',$$

$$\therefore \ \omega' = \tfrac{1}{2}\omega.$$

8. A rod of length a is moving about one end fixed with uniform angular velocity, when suddenly this end is freed, and a point distant l from it is fixed. What in general will be the direction and magnitude of the new angular velocity?

This is an extension of the preceding problem, and the method of solution will be similar. Let O (Fig. 39) be the first point fixed, and the angular velocity be ω, as indicated. Then this point being freed, let the second point O' be fixed.

The new angular velocity will be obtained by equating the moments of momentum before and after the fixing of the point O'. Thus

$$\rho \int_0^l \omega x(l-x) \cdot dx - \rho \int_0^{a-l} \omega x(l+x) \cdot dx = Mk^2 \text{ (about } O') \times \omega'.$$

For, the linear velocity of an element at P is $(l-x)\omega$ before O' is fixed, and its moment of momentum about O' will there-

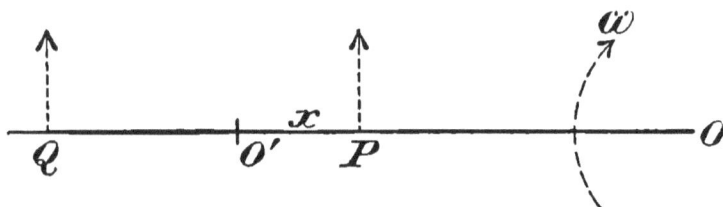

Fig. 39.

fore be $m(l-x)\omega \cdot x$; while the moment of momentum of an element at Q will be $m(l+x)\omega \cdot x$ in an opposite direction to the former with reference to the point O'. If ρ be the density and a the length of the rod, we then get the above relation which determines the sign and value of ω'.

It will be found on integrating the above expressions that ω' will have the same sign as ω, the opposite sign, or will be zero, according as

$$l > < = \tfrac{2}{3} a,$$

which shows that if a rod be moving about an axis, and this axis be freed and a new axis fixed through the centre of percussion, it will be reduced to rest.

9. An elliptic lamina is rotating with uniform angular velocity about one latus rectum, when suddenly the axis is freed and the other latus rectum fixed; find the new angular velocity.

$$\omega' = \frac{1 - 4\,\epsilon^2}{1 + 4\,\epsilon^2}\omega.$$

10. A circular plate rotates about an axis through its centre perpendicular to its plane with uniform angular velocity. If this axis be freed, and a point in the circumference of the plate be fixed, find the new angular velocity.

Fig. 40 gives the solution. For an element at P the linear velocity is $\omega \times OP$, and its moment of momentum about O' is $m\omega \times OP \times O'P$. If $OP = r$ and the radius of the plate be a, then will

$$\Sigma m\omega r(r - a \cos \theta) = Mk^2\omega'.$$

$$\therefore \rho\omega \int_0^a \int_0^{2\pi} r^2(r - a \cos \theta)\,dr\,d\theta = M \frac{3}{2}\frac{a^2}{2}\omega'.$$

$$\therefore \omega' = \tfrac{1}{3}\omega.$$

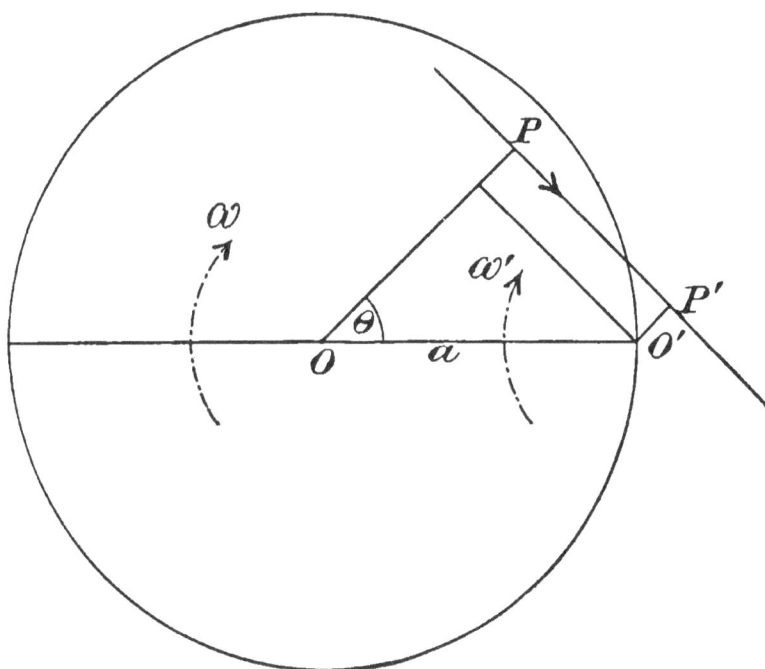

Fig. 40.

11. A circular plate is turning in its own plane about a point A on its circumference. Suddenly A is freed, and a

point B, also on the circumference, is fixed. Show that the plate will be reduced to rest if AB be one-third of the circumference.

12. A triangular plate ABC, right-angled at C, is rotating about AC. If AC be loosed suddenly, and BC fixed, find the new angular velocity.

$$\omega' = \frac{BC}{2\,AC}\omega.$$

13. A square lamina is rotating with angular velocity ω about a diagonal, when suddenly the diagonal is freed and one of the angular points not in the diagonal becomes fixed; prove that the angular velocity about this angular point will be $\frac{1}{7}\,\omega$.

14. A cube is rotating with angular velocity ω about a diagonal, when suddenly the diagonal is freed and one of the edges which does not meet that diagonal becomes fixed; prove that the angular velocity about this edge will be $\frac{1}{12}\,\omega\,\sqrt{3}$.

15. A uniform string hangs at rest over a smooth peg. If half the string on one side be cut off, show that the pressure on the peg is instantaneously reduced by one-third.

52. *The Ballistic Pendulum.*

This is a device for measuring the velocity of discharge of a rifle bullet, and was invented by Robins about 1743, and afterwards used by Dr. Hutton; and although of recent years superseded by the more accurate electric chronograph, it is to be noticed here as illustrating the nature of an impulse. In its simplest form it is a heavy pendulum capable of moving about a horizontal axis; a bullet discharged into it produces a certain angular velocity, and the pendulum rises through an angle which can be easily measured; or else a rifle is attached to it, and the discharge of the bullet produces a recoil.

The latter method is shown by Fig. 41, in which OA represents the pendulum, holding the rifle, and in its position of equilibrium under the action of gravity. The bullet being driven out produces a recoil through the angle α, and the velocity of discharge is found as follows :

Let m = mass of bullet,

v = its initial velocity,

l = distance of gun from O,

M = mass of pendulum and gun,

k = radius of gyration about O,

h = distance of centre of inertia from O.

$\therefore\ mvl = M(k^2 + h^2)\omega,$

where ω is the angular velocity generated.

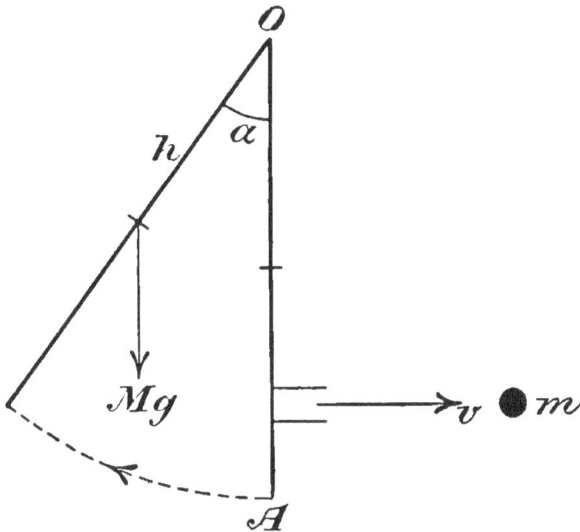

Fig. 41.

The pendulum then moves back through the angle α, which is observed, and its equation of motion on the way up is

$$\frac{d\omega}{dt} = -\frac{gh \sin\theta}{h^2 + k^2}.$$

$$\therefore \int_{\omega}^{0} 2\,\omega d\omega = -\frac{2\,gh}{h^2+k^2}\int_{0}^{\alpha} \sin\theta d\theta.$$

$$\therefore \omega^2 = \frac{2\,gh}{h^2+k^2}\,(1-\cos\alpha),$$

and this, combined with the previous relation, determines v.

In the other method a similar relation will be found; the only difference being that at each shot the pendulum is increased in mass by the addition of the bullet fired into it.

A rough pendulum made of a wooden box filled with sand, and attached to an iron bar which carries knife edges resting on a horizontal smooth plate, will readily illustrate the above equations.

The preceding solution assumes that the recoil of the pendulum, when the gun is fired without a ball, is so small that it may be neglected. Experiments have shown that this assumption may safely be made for small charges of powder but not for large charges. In the case of the latter, Hutton assumed that the effect of the charge of powder on the recoil is the same when the gun is fired with a ball as it is when it is fired without a ball. Consequently if the recoil is through an angle β when the gun is discharged without a ball, and through an angle α when it is discharged with a ball, the velocity of the ball will be

$$\frac{2\,M\sqrt{\{gh(k^2+h^2)\}}}{ml}\left(\sin\frac{\alpha}{2}-\sin\frac{\beta}{2}\right).$$

It has been found that the actual velocity of the ball lies between the velocities given by the two solutions.

CHAPTER VI.

Finite Forces.

53. If a body, fixed at one point only, moves under the action of any finite forces, then at every instant there is a line of particles at rest, so that the body is moving about what is called an *instantaneous axis* passing through the fixed point. Each particle will have a certain angular velocity about this axis, and the equations of motion with reference to any three rectangular axes passing through the fixed point can be written down in accordance with the principles already explained. In these equations the expressions for the effective forces will have to be evaluated in terms of the angular velocity about the instantaneous axis, and in order to explain how this may be done, the following propositions on the composition and resolution of angular velocities will be found useful.

54. *Angular velocity* is measured in the same manner as linear velocity : by the angle described in a unit of time if the motion be uniform, or by $\dfrac{d\theta}{dt}$ if the motion be not uniform. It may be represented by a straight line drawn in the proper direction, and perpendicular to the plane of rotation. And it will be seen that angular velocities can be compounded or resolved in the same way as forces acting at a point.

PROPOSITION I. — *For angular velocities about the same rotation axis, the resultant is the algebraic sum.*

This is evident, since the successive displacements in a small time are superimposed.

PROPOSITION 2. — *If a body have at any instant two angular velocities about two axes drawn from a point, and if lengths OA, OB be taken upon the axes to represent in direction and in magnitude the angular velocities, then the resultant angular velocity will be the diagonal OC of the parallelogram of which OA, OB are adjacent sides.*

Let a body, fixed at O, have two angular velocities represented in direction and in magnitude by OA, OB; and let the positive direction of rotation be with the hands of a watch. Take any point P in the plane containing OA, OB, and construct Fig. 42. And let $OA = \omega_a$, $OB = \omega_b$.

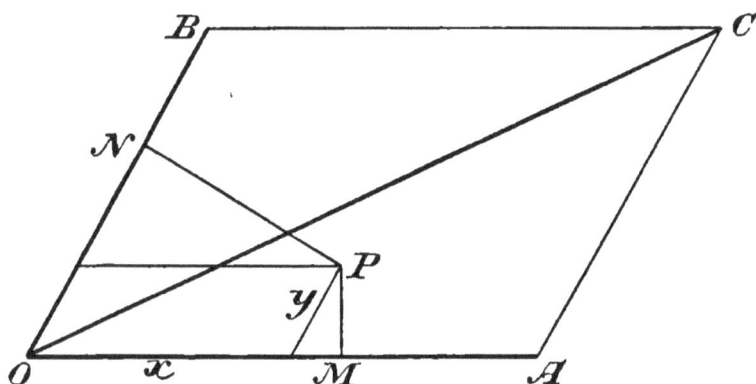

Fig. 42.

Then, owing to ω_a, the point P would be displaced downwards in an infinitely small time dt, a distance $\omega_a \cdot PM \cdot dt$ or $\omega_a y \sin AOB \cdot dt$. Due to ω_b its displacement would be upwards (above the plane of the paper) and equal to $\omega \, PNdt$ or $\omega_b x \sin AOB \cdot dt$.

Therefore the total displacement of P is

$$\sin AOB(y\omega_a - x\omega)dt,$$

and this is zero when

$$\frac{x}{\omega_a} = \frac{y}{\omega} \text{ or } \frac{x}{OA} = \frac{y}{OB},$$

which is the equation of the straight line OC. And thus for all points along OC there is no displacement; that is, the body is turning about OC, due to rotations about OA and OB.

That the line OC represents the magnitude of the resultant angular velocity may be shown by considering the displacement of the point A. Let ω_c be the resultant angular velocity about OC.

The displacement of A, due to ω_a, is zero.

The displacement of A, due to ω_b, is $OA \sin AOB \cdot \omega_b dt$.

The displacement of A, due to ω_c, is $OA \sin AOC \cdot \omega_c dt$, and therefore

$$OA \sin AOC \cdot \omega_c dt = OA \sin AOB \cdot \omega_b dt$$

$$\therefore \omega_c = OB \cdot \frac{\sin AOB}{\sin AOC} = OC.$$

PROPOSITION 3. — *If a body fixed at a point have angular velocities ω_x, ω_y, ω_z communicated to it about three rectangular axes passing through the fixed point, the resultant angular velocity is given by*

$$\omega^2 = \omega_x^2 + \omega_y^2 + \omega_z^2.$$

Also, if a body have an angular velocity ω about an instantaneous axis it may be said to have three angular velocities ω_x, ω_y, ω_z about three rectangular axes; and if α, β, γ be the angles which the instantaneous axis makes with the coördinate axes,

then

$$\frac{\omega_x}{\cos \alpha} = \frac{\omega_y}{\cos \beta} = \frac{\omega_z}{\cos \gamma} = \omega,$$

and

$$\frac{x}{\omega_x} = \frac{y}{\omega_y} = \frac{z}{\omega_z}$$

give the equations of the instantaneous axis when ω_x, ω_y, ω_z are known.

55. That a point may have at the same instant three angular velocities can be seen by means of the apparatus shown in Fig. 43.

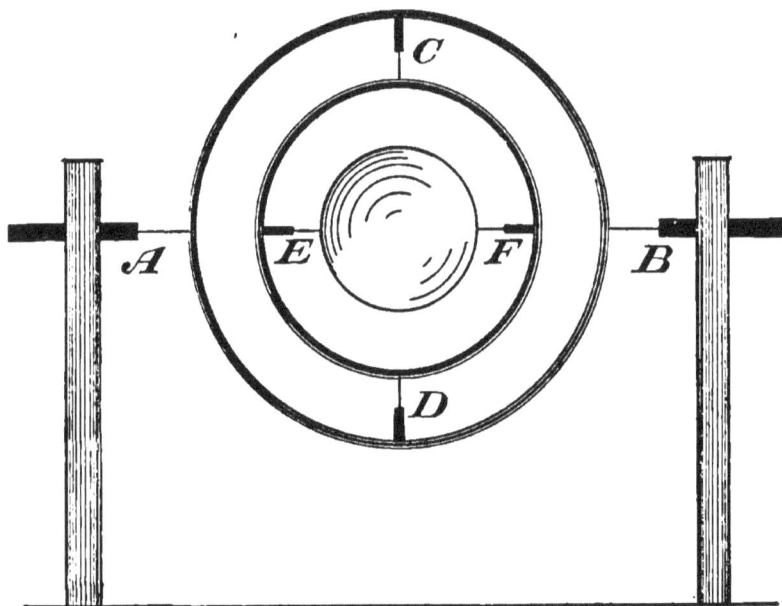

Fig. 43.

To an upright stand is attached by means of pivots a system of two rings and a sphere. The outer ring can rotate about an axis passing through the points A, B; the second ring may be made to rotate about CD; and the inner sphere about EF.

Now, the axis AB is initially in a horizontal position, and coincident with the axis of x drawn from O, the centre of the sphere; and if CD be made coincident with the axis of y by placing the plane of the outer ring in the plane of xy, then it is evident that by turning the inner ring the axis EF may be made initially coincident with the axis of z.

This having been done, rotations may be given first to the sphere, then to the inner ring, and lastly to the outer ring; and thus any point on the sphere will have simultaneously the three angular velocities given to the system, and the sphere will rotate about a resultant axis in space, which would be fixed were there no friction at the pivots and no resistance of the air.

The arrangement also shows how a heavy body may be fixed at its centre of gravity and at the same time be given rotations about axes fixed in space.

56. *Linear Velocity and Angular Velocity.*

In the case of a body moving with one point fixed we may replace the angular velocity ω about the instantaneous axis by ω_x, ω_y, ω_z about three rectangular axes drawn through the fixed point. The next thing to be done is to connect the expressions for the effective forces with these component angular velocities and the coördinates of any element of the body, and in order to do this we must obtain an expression for the linear velocities $\frac{dx}{dt}$, $\frac{dy}{dt}$, $\frac{dz}{dt}$ of any element at the point (x, y, z) in terms of x, y, z, and ω_x, ω_y, ω_z; on differentiating these expressions, we shall then obtain the linear accelerations. We may proceed either geometrically or by direct analysis.

1. *By Geometrical Displacement.*

Fig. 44 shows how the linear displacements arise from the rotations about the coördinate axes.

In the first figure the body is supposed to be fixed at O, and OI is the instantaneous axis about which the body is moving with angular velocity ω. The body may be supposed to have three rotations ω_x, ω_y, ω_z about the three coördinate axes instead of ω about the instantaneous axis. Then, considering positive rotations as those in the direction of the motion of the hands of a watch, and taking the displacements of the point $P(x, y, z)$ due to a rotation ω_x, we have, in the second figure, P moving along a small arc PQ in time dt, due to ω_x. This small displacement PQ is equivalent to two PR, RQ in the directions indicated. Hence we have,

$$PQ = \omega_x \cdot O'P \cdot dt,$$

$$\therefore PR = -PQ \cdot \frac{y}{O'P} = -y\omega_x dt,$$

and

$$QR = PQ \cdot \frac{x}{O'P} = x\omega_x dt.$$

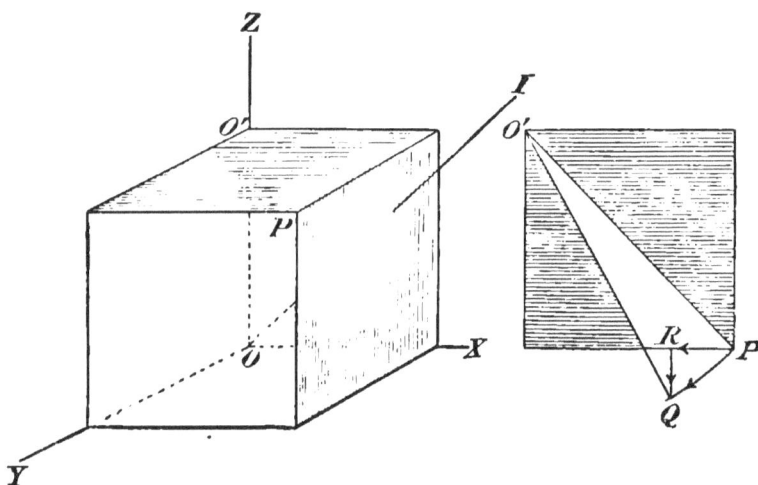

Fig. 44.

And, by considering the other planes, we should get the displacements of P due to ω_z and to ω_y, thus :

	Along Ox	Oy	Oz
Displacements due to ω_x	$-y\omega_x dt$	$x\omega_x dt$	
Displacements due to ω_z		$-z\omega_z dt$	$y_z\omega dt$
Displacements due to ω_y	$z\omega_y dt$		$-x\omega_y dt$

These are written down symmetrically ; and from them we see that the linear displacement along Ox, which we call dx, is equal to $(z\omega_y - y\omega_z)dt$, and, therefore, in the limit the linear velocity

$$\frac{dx}{dt} = z\omega_y - y\omega_z,$$

$$\frac{dy}{dt} = x\omega_z - z\omega_x,$$

and

$$\frac{dz}{dt} = y\omega_x - x\omega_y.$$

2. *By Direct Analysis.*

Let the body (Fig. 45) be fixed at the point O, and let OI be the instantaneous axis as before, and the angular velocity ω be

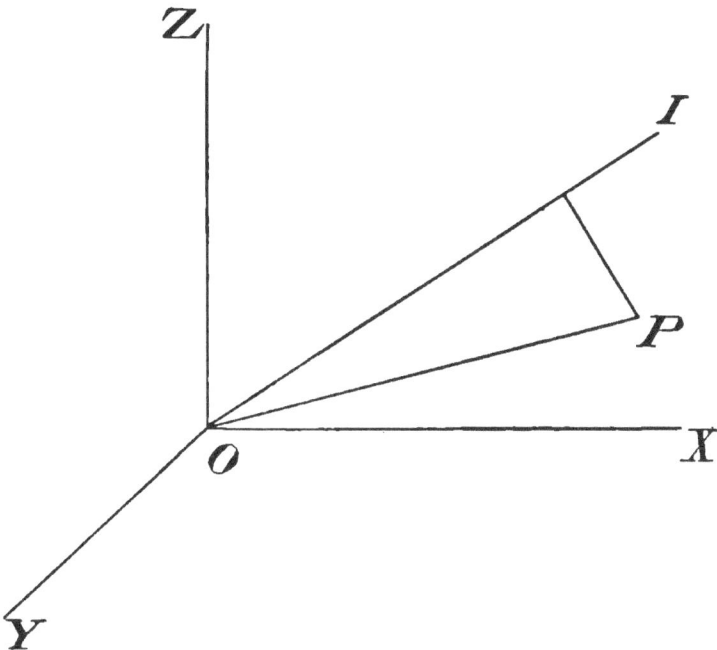

Fig. 45.

equivalent to ω_x, ω_y, ω_z, as shown. Then an element at P is tending to move at any instant in a circle about OI, and its absolute velocity is $\omega p = \dfrac{ds}{dt}$, where p is the perpendicular from P on the instantaneous axis.

And, if α, β, γ be the angles which OI makes with the coördinate axes, then

$$p^2 = (z \cos \beta - y \cos \gamma)^2 + \cdots + \cdots ;$$

also $\quad \dfrac{dx}{ds}, \dfrac{dy}{ds}, \dfrac{dz}{ds}$ are the direction cosines of the tangent at P,

$$\frac{x}{r}, \frac{y}{r}, \frac{z}{r} \text{ are the direction cosines of } OP,$$

$\cos \alpha$, $\cos \beta$, $\cos \gamma$ are the direction cosines of OI.

And, since OP is perpendicular to the tangent at P, and OI also perpendicular to this tangent, we have

$$\frac{dx}{ds} \cdot \frac{x}{r} + \frac{dy}{ds} \cdot \frac{y}{r} + \frac{dz}{ds} \cdot \frac{z}{r} = 0,$$

$$\frac{dx}{ds} \cos \alpha + \frac{dy}{ds} \cos \beta + \frac{dz}{ds} \cos \gamma = 0.$$

$$\therefore \frac{\dfrac{dx}{ds}}{z \cos \beta - y \cos \gamma} = \frac{\dfrac{dy}{ds}}{x \cos \gamma - z \cos \alpha} = \frac{\dfrac{dz}{ds}}{y \cos \alpha - x \cos \beta} = \frac{1}{p}.$$

And, therefore, since $\dfrac{ds}{dt} = \omega p$, we have, multiplying each quantity by $\dfrac{ds}{dt}$,

$$\frac{dx}{dt} = (z \cos \beta - y \cos \gamma)\omega = z\omega_y - y\omega_z,$$

$$\frac{dy}{dt} = \qquad \text{anal.} \qquad = x\omega_z - z\omega_x,$$

$$\frac{dz}{dt} = \qquad \text{anal.} \qquad = y\omega_x - x\omega_y,$$

as found before.

57. The former of the two investigations in the preceding article may be presented in purely analytical form thus :

(1) From the point P (Fig. 45) let fall perpendiculars on the coördinate axes OX, OY, OZ, and let θ, ϕ, ψ be the angles which these perpendiculars make with the coördinate planes XY, YZ, ZX. The angular velocity of P about the axis OX will be $\dfrac{d\theta}{dt}$, and the resolved parts of this parallel to the coördinate axes will be $\left(\dfrac{\partial x}{\partial \theta}\right)\dfrac{d\theta}{dt}$, $\left(\dfrac{\partial y}{\partial \theta}\right)\dfrac{d\theta}{dt}$ and $\left(\dfrac{\partial z}{\partial \theta}\right)\dfrac{d\theta}{dt}$ respectively.

Now
$$y = \sqrt{(r^2 - x^2)} \cos \theta,$$

$$z = \sqrt{(r^2 - x^2)} \sin \theta,$$

and
$$\frac{\partial x}{\partial \theta} = 0.$$

$$\therefore \frac{\partial y}{\partial \theta} = -\sqrt{(r^2 - x^2)} \sin \theta = -z,$$

and
$$\frac{\partial z}{\partial \theta} = \sqrt{(r^2 - x^2)} \cos \theta = y.$$

And by definition,
$$\frac{d\theta}{dt} = \omega_x.$$

$$\therefore \left(\frac{\partial y}{\partial \theta}\right)\frac{d\theta}{dt} = -z\omega_x,$$

and
$$\left(\frac{\partial z}{\partial \theta}\right)\frac{d\theta}{dt} = y\omega_x.$$

Treating the rotations about OY and OZ in like manner, we obtain the following complete system of equations :

$$x = \sqrt{(r^2 - z^2)} \cos \psi = \sqrt{(r^2 - y^2)} \sin \phi,$$

$$y = \sqrt{(r^2 - x^2)} \cos \theta = \sqrt{(r^2 - z^2)} \sin \psi,$$

$$z = \sqrt{(r^2 - y^2)} \cos \phi = \sqrt{(r^2 - x^2)} \sin \theta ;$$

$$\left(\frac{\partial x}{\partial \theta}\right)\frac{d\theta}{dt} = 0, \quad \left(\frac{\partial y}{\partial \theta}\right)\frac{d\theta}{dt} = -z\omega_x, \quad \left(\frac{\partial z}{\partial \theta}\right)\frac{d\theta}{dt} = y\omega_x,$$

$$\left(\frac{\partial y}{\partial \phi}\right)\frac{d\phi}{dt} = 0, \quad \left(\frac{\partial z}{\partial \phi}\right)\frac{d\phi}{dt} = -x\omega_y, \quad \left(\frac{\partial x}{\partial \phi}\right)\frac{d\phi}{dt} = z\omega_y,$$

$$\left(\frac{\partial z}{\partial \psi}\right)\frac{d\psi}{dt} = 0, \quad \left(\frac{\partial x}{\partial \psi}\right)\frac{d\psi}{dt} = -y\omega_z, \quad \left(\frac{\partial y}{\partial \psi}\right)\frac{d\psi}{dt} = x\omega_z.$$

The total velocity parallel to OX is the algebraic sum of the partial velocities, that is

$$\frac{dx}{dt} = \left(\frac{\partial x}{\partial \theta}\right)\frac{d\theta}{dt} + \left(\frac{\partial x}{\partial \phi}\right)\frac{d\phi}{dt} + \left(\frac{\partial x}{\partial \psi}\right)\frac{d\psi}{dt},$$

$$\therefore \frac{dx}{dt} = z\frac{d\phi}{dt} - y\frac{d\psi}{dt} = z\omega_y - y\omega_z.$$

Similarly,
$$\frac{dy}{dt} = x\frac{d\psi}{dt} - z\frac{d\theta}{dt} = x\omega_z - z\omega_x,$$

and
$$\frac{dz}{dt} = y\frac{d\theta}{dt} - x\frac{d\phi}{dt} = y\omega_x - x\omega_y.$$

(2) The second investigation in Art. 56 may also be presented in a purely analytical form thus:

$$x^2 + y^2 + z^2 = r^2, \tag{1}$$

$$x \cos \alpha + y \cos \beta + z \cos \gamma = r \cos \epsilon, \qquad (\epsilon = \text{angle } IOP),$$

$$p^2 = (z \cos \beta - y \cos \gamma)^2 + (x \cos \gamma - z \cos \alpha)^2 + (y \cos \alpha - x \cos \gamma)^2.$$

Also
$$\frac{\omega_x}{\cos \alpha} = \frac{\omega_y}{\cos \beta} = \frac{\omega_z}{\cos \gamma} = \omega,$$

and
$$\left(\frac{dx}{dt}\right)^2 + \left(\frac{dy}{dt}\right)^2 + \left(\frac{dz}{dt}\right)^2 = \left(\frac{ds}{dt}\right)^2 = p^2\omega^2.$$

$$\therefore x\omega_x + y\omega_y + z\omega_z = r\omega \cos \epsilon, \tag{2}$$

and
$$\left(\frac{dx}{dt}\right)^2 + \left(\frac{dy}{dt}\right)^2 + \left(\frac{dz}{dt}\right)^2 = (z\omega_y - y\omega_z)^2 + (x\omega_z - z\omega_x)^2 + (y\omega_x - x\omega_y)^2. \tag{3}$$

H

The body being rigid and O a fixed point in it, r and ϵ are constants; also ω, ω_x, ω_y, ω_z are independent of x, y, z, the coördinates of P, therefore from (1) and (2) we obtain, by differentiation,

$$x\frac{dx}{dt}+y\frac{dy}{dt}+z\frac{dz}{dt}=0,$$

$$\omega_z\frac{dx}{dt}+\omega_y\frac{dy}{dt}+\omega_z\frac{dz}{dt}=0.$$

$$\therefore\ \frac{\dfrac{dx}{dt}}{z\omega_y-y\omega_z}=\frac{\dfrac{dy}{dt}}{x\omega_z-z\omega_x}=\frac{\dfrac{dz}{dt}}{y\omega_x-x\omega_y}=\pm 1,\ \text{by (3).} \qquad (4)$$

The ambiguity of sign in (4) arises from the fact that in equations (2) and (3) there is nothing to determine whether the rotations ω_x, ω_y, ω_z, are in the directions x to y, y to z, z to x, or in the opposite directions x to z, z to y, y to x. If they are in the former directions, the value $+1$ must be taken, if they are in the latter directions, the value -1 must be taken.

58. *General Equations of Motion.*

Let the body be in motion, with one point O (Fig. 46) fixed, and let three rectangular axes be drawn from O, OX, OY, OZ, to which we may refer the position of the body at any time during its motion. And let it be acted upon by external forces, producing on each element of the body m, accelerations X, Y, Z in the directions of the three fixed axes. Then, if P be the pressure on the fixed point, and λ, μ, ν, the angles which the direction of the pressure makes with the fixed axes, we have, by D'Alembert's principle, the relations

$$\Sigma m\frac{d^2x}{dt^2}=\Sigma mX+P\cos\lambda, \qquad (1)$$

$$\Sigma m\frac{d^2y}{dt^2}=\Sigma mY+P\cos\mu, \qquad (2)$$

$$\Sigma m\frac{d^2z}{dt^2}=\Sigma mZ+P\cos\nu. \qquad (3)$$

And, also,

$$\Sigma m \left(y \frac{d^2z}{dt^2} - z \frac{d^2y}{dt^2} \right) = L, \tag{4}$$

$$\Sigma m \left(z \frac{d^2x}{dt^2} - x \frac{d^2z}{dt^2} \right) = M, \tag{5}$$

$$\Sigma m \left(x \frac{d^2y}{dt^2} - y \frac{d^2x}{dt^2} \right) = N, \tag{6}$$

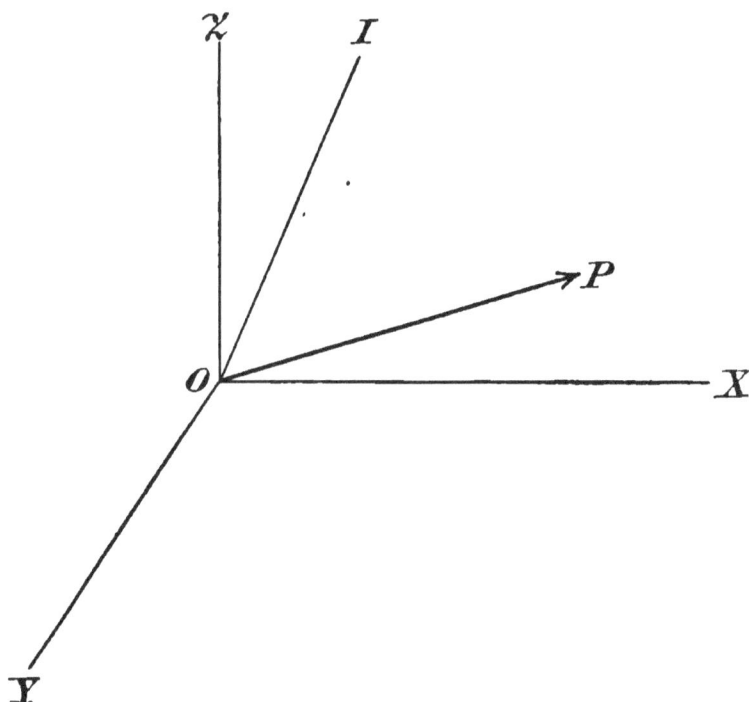

Fig. 46.

where L, M, N are the couples due to the external forces. Now, since the body, when we form the above equations, is moving about an instantaneous axis OI with some angular velocity ω, which we may suppose equivalent to ω_x, ω_y, ω_z, and since we have shown that $\frac{dx}{dt} = z\omega_y - y\omega_z$, $\frac{dy}{dt} = x\omega_z - z\omega_x$, and $\frac{dz}{dt} = y\omega_x - x\omega_y$, it is evident that these equations can be expressed in terms of known quantities, and ω_x, ω_y, ω_z.

Hence the equations (4), (5), (6), taken with the initial circumstances, will serve to determine ω_x, ω_y, ω_z, and therefore ω and the position of the instantaneous axis; equations (1), (2), (3) will then give, on substitution, the value of P.

59. *Equations of Motion referred to Axes fixed in Space.*

Taking the equation (6), we shall proceed to evaluate it in terms of ω_x, ω_y, ω_z by taking the values

$$\frac{dx}{dt} = z\omega_y - y\omega_z, \quad \frac{dy}{dt} = x\omega_z - z\omega_x, \quad \frac{dz}{dt} = y\omega_x - x\omega_y,$$

and differentiating.

Thus we should get $\dfrac{d^2x}{dt^2} = z\dfrac{d\omega_y}{dt} + \omega_y\dfrac{dz}{dt} - y\dfrac{d\omega_z}{dt} - \omega_z\dfrac{dy}{dt}$, and on substituting in this the values of $\dfrac{dz}{dt}$, $\dfrac{dy}{dt}$, we get

$$\frac{d^2x}{dt^2} = z\frac{d\omega_y}{dt} - y\frac{d\omega_z}{dt} - x(\omega_x^2 + \omega_y^2 + \omega_z^2) + \omega_x(x\omega_x + y\omega_y + z\omega_z).$$

$$\frac{d^2x}{dt^2} = z\frac{d\omega_y}{dt} - y\frac{d\omega_z}{dt} - \omega^2 x + \omega_x(x\omega_x + y\omega_y + z\omega_z),$$

and, similarly, we should get, by symmetry,

$$\frac{d^2y}{dt^2} = x\frac{d\omega_z}{dt} - z\frac{d\omega_x}{dt} - \omega^2 y + \omega_y(x\omega_x + y\omega_y + z\omega_z).$$

Therefore relation (6) becomes

$$N = \Sigma m\left(x\frac{d^2y}{dt^2} - y\frac{d^2x}{dt^2}\right)$$

$$= \Sigma m(x^2 + y^2) \cdot \frac{d\omega_z}{dt} - \Sigma mxz\frac{d\omega_x}{dt} - \Sigma myz\frac{d\omega_y}{dt}$$

$$+ \Sigma m(x\omega_x + y\omega_y + z\omega_z)(x\omega_y - y\omega_x),$$

and by analogy M and L can be written down.   Since L, M, N are given, these results would give the values of ω_x, ω_y, ω_z on integration, after calculation of the moments and products of inertia required.   But this calculation, as can be seen, would be tedious, and we can avoid most of it by choosing axes which although still fixed in space are in coincidence with the principal axes of the body when we form the equations of motion. This device enables us at once to disregard the products of inertia, and makes a great simplification in the problem.   It is due to *Euler*, and the equations thus obtained are known as *Euler's* equations of motion.

60. *Euler's Equations of Motion.*

Instead of choosing any three rectangular axes fixed in space at the instant under consideration, let axes be so chosen that they coincide with the principal axes of the moving body; and let ω_1, ω_2, ω_3 be the angular velocities about these principal axes, which will then be the same as ω_x, ω_y, ω_z in the preceding equations.   We shall then have

$$N = \Sigma m(x^2 + y^2)\frac{d\omega_z}{dt} + \Sigma m(x^2 - y^2)\omega_1\omega_2,$$

and it can be shown (see Art. 61), that $\dfrac{d\omega_z}{dt} = \dfrac{d\omega_3}{dt}$.

$$\therefore\ N = \Sigma m(x^2 + y^2)\frac{d\omega_3}{dt} + \Sigma m(x^2 - y^2)\omega_1\omega_2.$$

Thus the equations for determining ω_1, ω_2, ω_3 become

$$A\frac{d\omega_1}{dt} - (B - C)\omega_2\omega_3 = L,$$

$$B\frac{d\omega_2}{dt} - (C - A)\omega_3\omega_1 = M,$$

$$C\frac{d\omega_3}{dt} - (A - B)\omega_1\omega_2 = N.$$

These equations, on integration, being three in number, should serve theoretically to determine the angular velocities ω_1, ω_2, ω_3, and the position of the instantaneous axis. The actual situation of the body with reference to known directions in space can also be found from these, combined with certain other relations which will be given further on.

61. It might seem that $\dfrac{d\omega_2}{dt} = \dfrac{d\omega_3}{dt}$ follows at once from the relation $\omega_2 = \omega_3$, but it does not necessarily so follow; that the

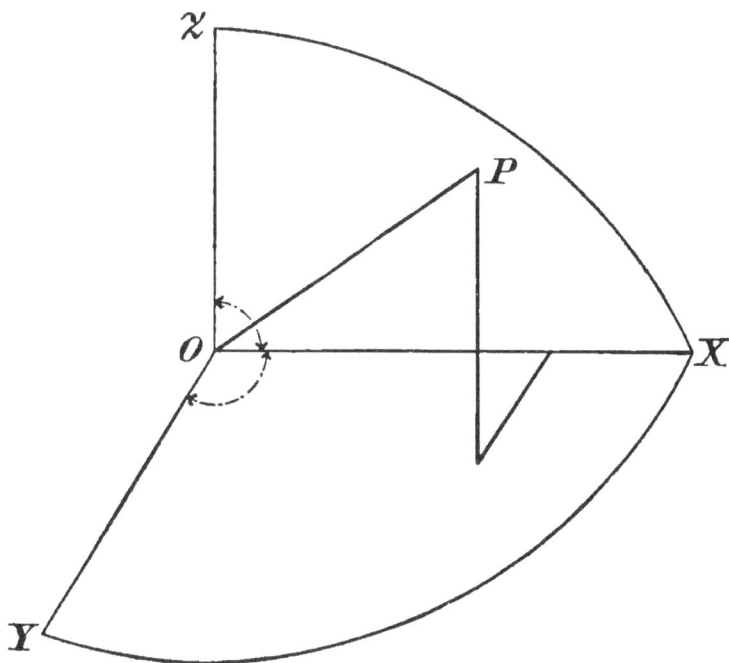

Fig. 47.

former relation holds as well as the latter may however be shown in the following manner:

Let OX, OY, OZ be the three axes fixed in space (Fig. 47);

then a body moving with the point O fixed will produce about OP an angular velocity

$$\omega_x \cos \alpha + \omega_y \cos \beta + \omega_z \cos \gamma,$$

if α, β, γ are the angles which OP makes with the axes.

Differentiating this expression with respect to t, we get for the angular acceleration about OP

$$\cos \alpha \frac{d\omega_x}{dt} - \omega_x \sin \alpha \frac{d\alpha}{dt} + \cos \beta \frac{d\omega_y}{dt} - \omega_y \sin \beta \frac{d\beta}{dt} + \cos \gamma \frac{d\omega_z}{dt}$$

$$- \omega_z \sin \gamma \frac{d\gamma}{dt}.$$

Suppose now that OP approaches OX and ultimately coincides with it, then the angular acceleration becomes

$$\frac{d\omega_x}{dt} - \omega_y \frac{d\beta}{dt} - \omega_z \frac{d\gamma}{dt},$$

because $\alpha = 0$, $\beta = \gamma = \dfrac{\pi}{2}$ when OP coincides with OX; and it is also evident from the figure that in such case $\dfrac{d\beta}{dt}$ is the same as ω_z or ω_3, and that $\dfrac{d\gamma}{dt} = -\omega_y$ or $-\omega_2$.

$$\therefore \frac{d\omega_1}{dt} = \frac{d\omega_x}{dt} \text{ at the same time that } \omega_1 = \omega_x.$$

The relations between $\dfrac{d\omega_1}{dt}$, $\dfrac{d\omega_2}{dt}$, $\dfrac{d\omega_3}{dt}$, the angular accelerations around axes fixed in the body, and $\dfrac{d\omega_x}{dt}$, $\dfrac{d\omega_y}{dt}$, $\dfrac{d\omega_z}{dt}$, the angular accelerations around axes fixed in space, may be determined for any given position of the moving body, as follows:

Let l_1, m_1, n_1; l_2, m_2, n_2; l_3, m_3, n_3, be the direction cosines of axes fixed in the body referred to coördinate axes fixed in space. Then will

$$\left.\begin{aligned}
\omega_1 &= l_1\omega_x + m_1\omega_y + n_1\omega_z \\
\omega_2 &= l_2\omega_x + m_2\omega_y + n_2\omega_z \\
\omega_3 &= l_3\omega_x + m_3\omega_y + n_3\omega_z
\end{aligned}\right\} \quad (1)$$

$$\left.\begin{aligned}
\omega_x &= l_1\omega_1 + l_2\omega_2 + l_3\omega_3 \\
\omega_y &= m_1\omega_1 + m_2\omega_2 + m_3\omega_3 \\
\omega_z &= n_1\omega_1 + n_2\omega_2 + n_3\omega_3
\end{aligned}\right\} \quad (2)$$

$$\left.\begin{aligned}
l_1^2 + m_1^2 + n_1^2 &= 1, & l_1l_2 + m_1m_2 + n_1n_2 &= 0, \\
l_2^2 + m_2^2 + n_2^2 &= 1, & l_2l_3 + m_2m_3 + n_2n_3 &= 0, \\
l_3^2 + m_3^2 + n_3^2 &= 1, & l_3l_1 + m_3m_1 + n_3n_1 &= 0.
\end{aligned}\right\} \quad (3)$$

Differentiating the first equation in group (1),

$$\frac{d\omega_1}{dt} = l_1\frac{d\omega_x}{dt} + m_1\frac{d\omega_y}{dt} + n_1\frac{d\omega_z}{dt} + \omega_x\frac{dl_1}{dt} + \omega_y\frac{dm_1}{dt} + \omega_z\frac{dn_1}{dt}.$$

But the sum of the last three terms on the right hand side of this equation is zero, for

$$\omega_x\frac{dl_1}{dt} + \omega_y\frac{dm_1}{dt} + \omega_z\frac{dn_1}{dt}$$

$$= (l_1\omega_1 + l_2\omega_2 + l_3\omega_3)\frac{dl_1}{dt} + (m_1\omega_1 + m_2\omega_2 + m_3\omega_3)\frac{dm_1}{dt}$$

$$+ (n_1\omega_1 + n_2\omega_2 + n_3\omega_3)\frac{dn_1}{dt}$$

$$= \left(l_1\frac{dl_1}{dt} + m_1\frac{dm_1}{dt} + n_1\frac{dn_1}{dt}\right)\omega_1 + \left(l_2\frac{dl_1}{dt} + m_2\frac{dm_1}{dt} + n_2\frac{dn_1}{dt}\right)\omega_2$$

$$+ \left(l_3\frac{dl_1}{dt} + m_3\frac{dm_1}{dt} + n_3\frac{dn_1}{dt}\right)\omega_3$$

$$= 0.$$

as appears at once on differentiating the equations in group (3).

$$\therefore \frac{d\omega_1}{dt} = l_1\frac{d\omega_x}{dt} + m_1\frac{d\omega_y}{dt} + n_1\frac{d\omega_z}{dt}. \tag{4}$$

From the second and third equations in group (1) we may in like manner obtain

$$\frac{d\omega_2}{dt} = l_2\frac{d\omega_x}{dt} + m_2\frac{d\omega_y}{dt} + n_2\frac{d\omega_z}{dt},$$

and

$$\frac{d\omega_3}{dt} = l_3\frac{d\omega_x}{dt} + m_3\frac{d\omega_y}{dt} + n_3\frac{d\omega_z}{dt}.$$

Hence the acceleration around any axis may be projected on coördinate axes just as angular velocities and as segments of the rotation axis may be projected, and all theorems on the projection of segments of a line may be interpreted as theorems on the projection of angular accelerations about the line.

If the axis of ω_1 coincide at any moment with the axis of ω_z, then will $l_1 = 1$, $m_1 = 0$, $n_1 = 0$, $\omega_1 = \omega_z$, and by (4) above

$$\frac{d\omega_1}{dt} = \frac{d\omega_z}{dt}.$$

62. *Angular Coördinates of the Body.*

The equations of motion known as Euler's enable us to find ω_1, ω_2, ω_3, the angular velocities of the body with reference to the principal axis drawn through the fixed point about which the body is moving. As these principal axes, however, are in the body, and move with it, we must have some means of determining the *position* of the body with reference to axes *fixed in space*, because the values of the angular velocities found by solving Euler's equations tell us nothing whatever as yet of the situation of the body with regard to any known directions in space. In order, then, to fix the position of the body at any time and give us a definite idea of its situation with reference to some initial position, three angles θ, ϕ, ψ

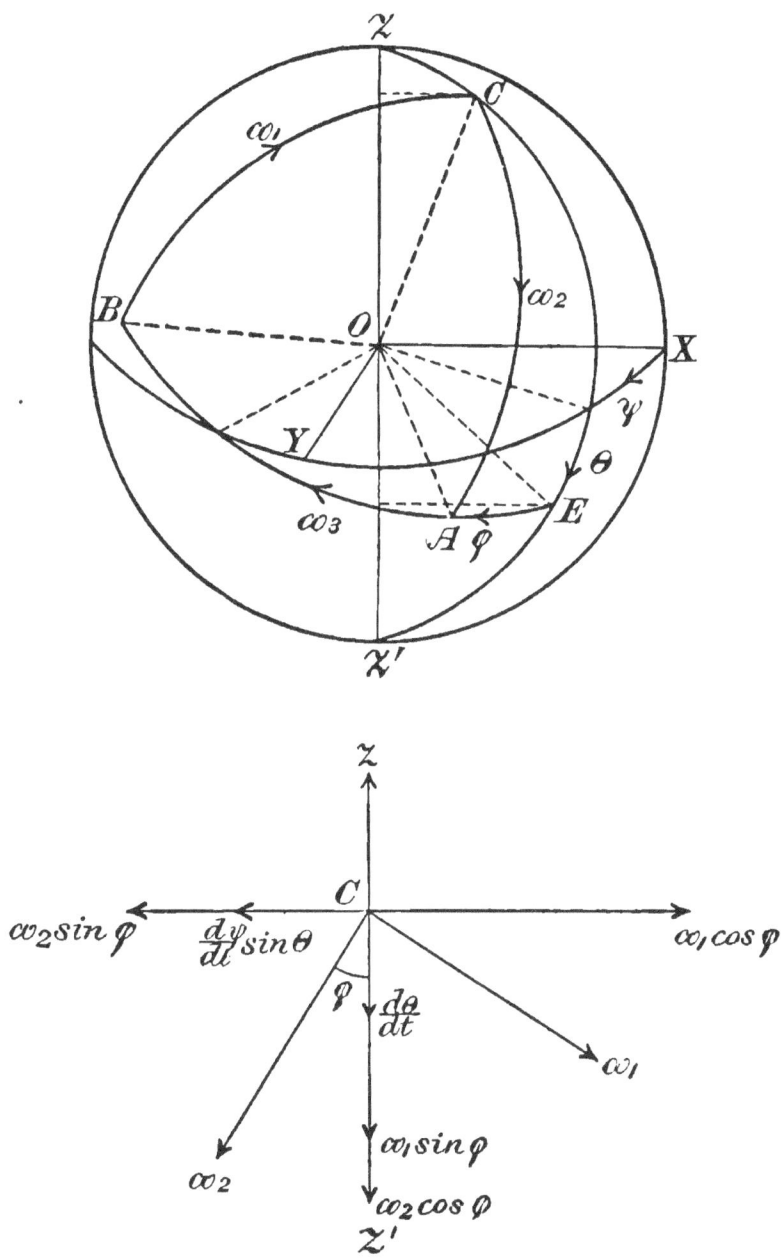

Fig. 48.

are chosen, known as the angular coördinates ; they define the
situation of the principal axes, and therefore of the body itself,
being measured from some initial fixed axes of reference which,
at the beginning of the motion, coincide with the principal axes
of the body. Relations can be easily found between θ, ϕ, ψ,
and ω_1, ω_2, ω_3, so that knowing the angular velocities we can
find θ, ϕ, ψ, and the motion of the body is fully known. The
subjoined figures (Fig. 48 and Fig. 49) show how the position
of the principal axes at any instant may be determined by
displacements θ, ϕ, ψ ; they also indicate how the relations
existing between these displacements and the angular velocities
about the principal axes are to be found.

Let a spherical surface of radius unity be constructed at the
fixed point O (Fig. 48), about which we suppose a body to be
moving. Initially, let the body, which we may represent by its
principal axes OA, OB, OC, be in such a position that OA, OB,
OC coincide with OX, OY, OZ respectively. Then, by supposing the body to turn through the angles ψ, θ, ϕ in order, so
that the point A travels in the directions indicated by the
arrows, it is evident that *any* position of the body will be fully
known in respect of the fixed axes OX, OY, OZ, when we know
three such angles as θ, ϕ, ψ.

At any instant the body has angular velocities ω_1, ω_2, ω_3 indicated by arrows ; and in order to connect these with the angular
coördinates, consider the motion of a particular point such as C.
The velocity of the point C at the instant in question, may be
considered as the resultant of the angular velocities ω_1, ω_2, ω_3,
or as due to changes in θ, ϕ, ψ, *i.e.*, to velocities $\dfrac{d\theta}{dt}$, $\dfrac{d\phi}{dt}$, $\dfrac{d\psi}{dt}$;
and by expressing in the two systems of change the velocity of
C resolved in three determinate directions, and equating the
results, we shall arrive at the relations between ω_1, ω_2, ω_3, and
$\dfrac{d\theta}{dt}$, $\dfrac{d\phi}{dt}$, $\dfrac{d\psi}{dt}$.

The auxiliary figure shows the motion of the point C due to the two systems. The line ZCZ' is the tangent to the line of the great circle, and the point C will evidently have angular velocities ω_1, ω_2 in the directions indicated by the arrows; it has also a motion $\dfrac{d\theta}{dt}$ along the tangent to the great circle at C, and a motion $\dfrac{d\psi}{dt} \sin \theta$ perpendicular to this former. This velocity $\dfrac{d\psi}{dt} \sin \theta$ arises from the fact that C, owing to the ψ motion, has a velocity along a tangent to a small circle with CC' as radius, and its velocity perpendicular to ZCZ' must be $CC' \cdot \dfrac{d\psi}{dt} = OC \sin \theta \cdot \dfrac{d\psi}{dt} = \dfrac{d\psi}{dt} \sin \theta$, since we have agreed to call the radius OC unity.

Hence, we have from the auxiliary figure, remembering that the radius is unity, the relations,

$$\text{velocity of } C \text{ along } ZC = \frac{d\theta}{dt} = \omega_1 \sin \phi + \omega_2 \cos \phi, \tag{1}$$

$$\text{velocity of } C \text{ perpendicular to } ZC = \frac{d\psi}{dt} \sin \theta$$
$$= -\omega_1 \cos \phi + \omega_2 \sin \phi. \tag{2}$$

And by considering the motion of the point E, we have the velocity of E along the tangent at E equal to

$$\frac{d\phi}{dt} + OE \cos \theta \cdot \frac{d\psi}{dt} = \frac{d\phi}{dt} + \frac{d\psi}{dt} \cos \theta = \omega_3. \tag{3}$$

The relations (1), (2), (3), along with Euler's equations of motion, Art. 60, give a complete solution of the problem as far as the actual motion and position of the body are concerned.

Fig. 49 is given merely to show how the principal axes which at any time really represent the body itself were initially coincident with the fixed axes in space, and have turned through angles θ, ϕ, ψ. The complications in the former figure are omitted.

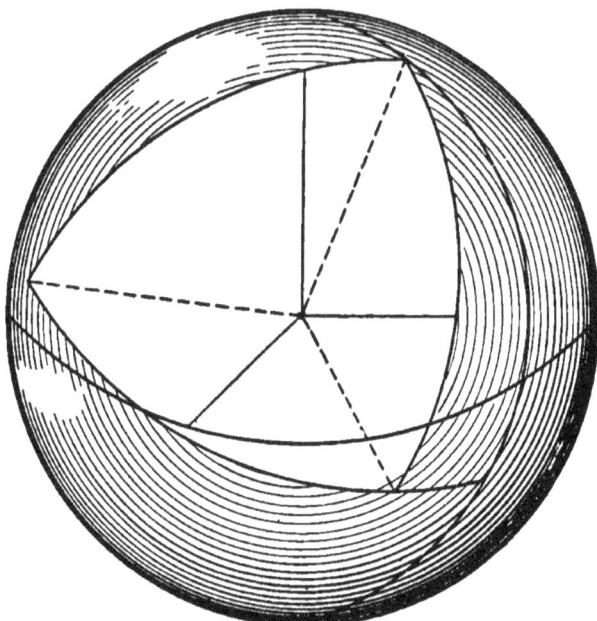

Fig. 49.

63. *Pressure on the Fixed Point.*

The pressures on the fixed point, measured along three fixed rectangular axes, will be given by the equations,

$$\Sigma m \frac{d^2x}{dt^2} = \Sigma m X + P \cos \lambda,$$

$$\Sigma m \frac{d^2y}{dt^2} = \Sigma m Y + P \cos \mu,$$

$$\Sigma m \frac{d^2z}{dt^2} = \Sigma m Z + P \cos \nu,$$

where $\Sigma m \dfrac{d^2x}{dt^2}$ is now to be expressed in terms of the coördinates of the centre of inertia, the mass of the body, and the angular velocities.   Thus, if we evaluate as formerly $\dfrac{d^2x}{dt^2}$ in terms of ω_x, ω_y, ω_z, we get

$$\Sigma m \frac{d^2x}{dt^2} = \Sigma m \left\{ z\frac{d\omega_y}{dt} - y\frac{d\omega_z}{dt} - \omega^2 x + \omega_x(x\omega_x + y\omega_y + z\omega_z) \right\};$$

and if $\bar{x}$, $\bar{y}$, $\bar{z}$ be the coördinates of the centre of inertia, we have, on reduction, to determine the three pressures,

$$\text{Mass} \cdot \left\{ \bar{z}\frac{d\omega_y}{dt} - \bar{y}\frac{d\omega_z}{dt} - \omega^2\bar{x} + \omega_x(\bar{x}\omega_x + \bar{y}\omega_y + \bar{z}\omega_z) \right\} = P\cos\lambda + \Sigma mX,$$

and two similar relations for $P\cos\mu$, $P\cos\nu$.

These equations are with reference to axes fixed in space; but if we refer them to the principal axes moving with the body, we may use Euler's equations, and substitute for $\dfrac{d\omega_y}{dt}$, $\dfrac{d\omega_z}{dt}$, $\dfrac{d\omega_z}{dt}$ their values in terms of A, B, C, L, M, N, ω_1, ω_2, ω_3.

The equations when finally reduced in this way become

$$\text{Mass} \cdot \left\{ \omega_1(B + C - A)\left(\frac{\bar{y}\omega_2}{C} + \frac{\bar{z}\omega_3}{B}\right) - (\omega_2^2 + \omega_3^2)\bar{x} \right\}$$

$$= P\cos\lambda + \Sigma mX - \text{Mass} \cdot \left\{ \frac{M}{B}\bar{z} - \frac{N}{C}\bar{y} \right\},$$

with the two analogous expressions for $P\cos\mu$, $P\cos\nu$.   In these expressions $\bar{x}$, $\bar{y}$, $\bar{z}$ are the coördinates of the centre of inertia, L, M, N the couples due to the external forces, A, B, C the principal moments at the fixed point.

And it is evident that if $\bar{x}=\bar{y}=\bar{z}=0$, the pressure on the fixed point will be the resultant of the external forces ΣmX, ΣmY, ΣmZ; as, for example, in the case of a heavy body fixed at its centre of gravity, where the pressure must be simply the weight of the body.

Illustrative Examples.

1. If ω_x, ω_y, ω_z be the angular velocities about the coördinate axes by which the motion of a body about the origin may be exhibited, find the locus of the points the magnitude of whose velocity is $a\omega_x$.

2. The locus of points in a body (which is moving with one point fixed) that have at any proposed instant velocities of the same magnitude, is a circular cylinder.

3. A body fixed at one point moves so that its angular velocities about its principal axes are $a \sin nt$, $a \cos nt$, in which t represents the time, and n and a are constants. Show that the instantaneous axis describes a circular cone in the body with uniform velocity.

4. A uniform rod, length $2\,a$, turns freely about its upper end, which is fixed, and revolves so as to be constantly inclined at an angle α to the vertical. Find the direction and magnitude of the pressure on the fixed end.

5. Any heavy body, for which the momental ellipsoid at the centre of inertia is a sphere, will, if fixed at its centre of inertia, continue to revolve about any axis around which it was originally put in motion.

6. A right circular cone, whose altitude is equal to the diameter of its base, turns about its centre of inertia, which is fixed, and is originally put in motion about an axis inclined at an angle α to its axis of figure. Show that the vertex of the cone will describe a circle whose radius is $\frac{3}{4} a \sin \alpha$, a being the altitude.

This is evident, since the momental ellipsoid at the centre of inertia of the cone is a sphere; therefore the cone will revolve about the original axis permanently (Ex. 5 above), and its axis will describe another cone, and its apex will trace out a circle of radius $\frac{3}{4}\,a\sin\alpha$.

7. A circular plate revolves about its centre of gravity fixed. If an angular velocity ω were originally impressed upon it about an axis making an angle α with its plane, show that a normal to the plane of the plate will make a revolution in space in time

$$\frac{2\,\pi}{\omega\sqrt{1+3\sin^2\alpha}}.$$

8. A body has an angular velocity ω about a line passing through the point α, β, γ, and having direction cosines l, m, n. Show that the motion is equivalent to rotations $l\omega$, $m\omega$, $n\omega$ about the coördinate axes and translations $(m\gamma-n\beta)\omega$, $(n\alpha-l\gamma)\omega$, $(l\beta-m\alpha)\omega$ in the directions of these axes.

9. A body has equal angular velocities about two axes which neither meet nor are parallel. Show that the motion is equivalent to a translation along a line equally inclined to the two axes and a rotation about this line.

64. *Top spinning on a Rough Horizontal Plane.*

When a common top, symmetrical with respect to its axis, is spun and placed on a rough horizontal plane, with its axis inclined at an angle to the vertical, it satisfies approximately the conditions for motion about a fixed point; and we may first consider the ideal case of a top, spinning on a perfectly rough horizontal plane, with its apex fixed, and free to move in all directions about this apex considered as a fixed point.

Let a top, Fig. 50 (1), be set spinning about its axis, and placed on a rough horizontal plane, with its axis inclined to the vertical at a given angle. Then after a certain time its position with reference to fixed lines in space will be as indicated in the figure by its principal axes OA, OB, OC, drawn through the fixed point. G is the centre of gravity of the top, and $OG=h$.

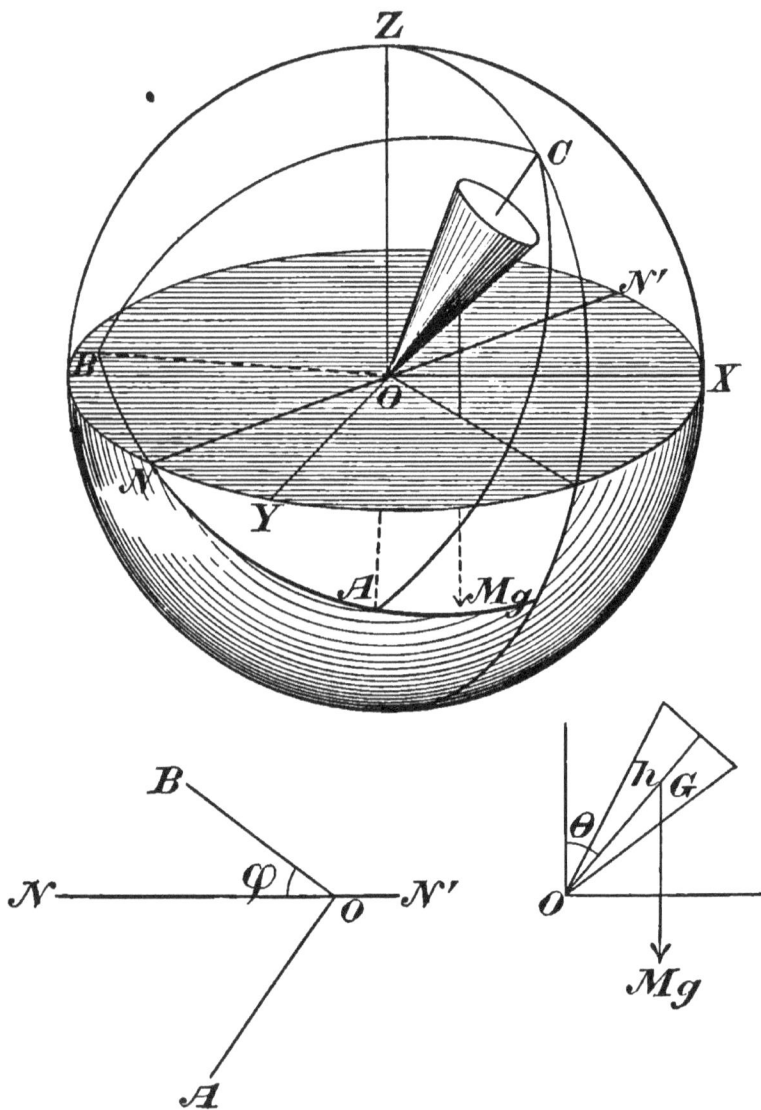

Fig. 50.

The angle $ZOC = \theta$, and the line NON' is the *line of nodes*. The angular velocity $\dfrac{d\theta}{dt}$ is called the *Nutation*, and $\dfrac{d\psi}{dt}$ the *Precession*.

The top is acted upon only by the external force of gravity, since we suppose an ideal case first and neglect the couple of friction acting at the fixed point, as well as the resistance of the air. The external couple is equal to $Mgh \sin \theta$, as is seen from Fig. 50 (3), which tends to turn the top about the line of nodes. This couple may evidently be resolved into two others, one equal to $Mgh \sin \theta \cos \phi$, tending to turn the top about OB, and the other, equal to $Mgh \sin \theta \sin \phi$, tending to turn it about OA, as may be seen from Fig. 50 (2). Hence the equations of motion are

$$A\frac{d\omega_1}{dt} - (B-C)\omega_2\omega_3 = Mgh \sin \theta \sin \phi, \qquad (1)$$

$$B\frac{d\omega_2}{dt} - (C-A)\omega_3\omega_1 = Mgh \sin \theta \cos \phi, \qquad (2)$$

$$C\frac{d\omega_3}{dt} - (A-B)\omega_1\omega_2 = 0, \qquad (3)$$

and we have also the relations

$$\frac{d\theta}{dt} = \omega_1 \sin \phi + \omega_2 \cos \phi, \qquad (4)$$

$$\frac{d\psi}{dt} \sin \theta = \omega_2 \sin \phi - \omega_1 \cos \phi, \qquad (5)$$

$$\frac{d\phi}{dt} + \frac{d\psi}{dt} \cos \theta = \omega_3, \qquad (6)$$

and it is known that $A = B$, since the top is symmetrical about its axis; and that ω_3 has an initial value n given to it in spinning, while θ has an initial value θ_0, at which inclination to the vertical we place the top at the beginning of the motion.

Equation (3) becomes

$$C\frac{d\omega_3}{dt}=0.$$

$$\therefore \; \omega_3 = \text{constant} = n, \text{ its initial value.}$$

Equations (1) and (2) give, when multiplied by ω_1, ω_2, respectively, and added together,

$$A\left(\omega_1\frac{d\omega_1}{dt}+\omega_2\frac{d\omega_2}{dt}\right)=Mgh\sin\theta(\omega_1\sin\phi+\omega_2\cos\phi),$$

which, by aid of (4), becomes

$$A\left(\omega_1\frac{d\omega_1}{dt}+\omega_2\frac{d\omega_2}{dt}\right)=Mgh\sin\theta\frac{d\theta}{dt}.$$

This, on integration, gives

$$A\int_0^{\omega_1}2\,\omega_1 d\omega_1+A\int^{\omega_2}2\,\omega_2 d\omega_2=2\,Mgh\int_{\theta_0}^{\theta}\sin\theta d\theta.$$

$$\therefore \; A(\omega_1{}^2+\omega_2{}^2)=2\,Mgh(\cos\theta_0-\cos\theta).$$

But, taking (4) and (5), and squaring and adding, we get

$$\omega_1{}^2+\omega_2{}^2=\left(\frac{d\theta}{dt}\right)^2+\left(\frac{d\psi}{dt}\right)^2\sin^2\theta.$$

$$\therefore \; A\left(\frac{d\theta}{dt}\right)^2+A\sin^2\theta\left(\frac{d\psi}{dt}\right)^2=2\,Mgh(\cos\theta_0-\cos\theta), \qquad (a)$$

which, as will be seen hereafter, is one form of the equation of energy.

65. In order to obtain another relation between $\dfrac{d\theta}{dt}$, $\dfrac{d\psi}{dt}$, we may proceed as follows.

Multiply (1) by $\cos\phi$ and (2) by $\sin\phi$ and subtract, and we get

$$\sin\phi\frac{d\omega_2}{dt}-\cos\phi\frac{d\omega_1}{dt}=\frac{C-A}{A}n\frac{d\theta}{dt}, \qquad \text{from (4).}$$

But by (5),

$$\frac{d\psi}{dt}\sin\theta = \omega_2\sin\phi - \omega_1\cos\phi.$$

$$\therefore \frac{d\psi}{dt}\cos\theta\frac{d\theta}{dt} + \sin\theta\frac{d^2\psi}{dt^2} = \frac{d\theta}{dt}\left(\frac{d\phi}{dt} + \frac{C-A}{A}n\right),$$

and since by (6)

$$\frac{d\phi}{dt} = n - \frac{d\psi}{dt}\cos\theta,$$

$$\therefore 2\frac{d\psi}{dt}\cos\theta\frac{d\theta}{dt} + \sin\theta\frac{d^2\psi}{dt^2} = \frac{Cn}{A}\cdot\frac{d\theta}{dt},$$

and therefore, multiplying through by $\sin\theta$ and integrating, we have

$$A\sin^2\theta\frac{d\psi}{dt} = Cn(\cos\theta_0 - \cos\theta), \qquad (b)$$

since, initially, the value of the left-hand side of the relation is zero.

66. The relation (b), which may also be obtained geometrically, shows that the sign of $\frac{d\psi}{dt}$ depends upon the sign of $n(\cos\theta_0 - \cos\theta)$, since A and C are positive quantities. In the case we have supposed, $\cos\theta_0$ is always greater than $\cos\theta$, since θ_0 is the least angle the axis of the top can make with the vertical; if the top were spun so that the centre of gravity were below the fixed point, then $\cos\theta_0 < \cos\theta$. Thus we see that the signs of $\frac{d\psi}{dt}$ and of ω_3 will be the same or opposite, according as the centre of gravity is above or below the fixed point. This is equivalent to saying that the motion of precession which the top acquires is direct or retrograde according as the centre of gravity is above or below the fixed apex about which it moves.

This motion of precession and its sign can easily be shown by a top of special construction, which is so arranged that one can alter at will the position of its centre of gravity.

A section of the top is shown in Fig. 51. It consists of an axis of steel *AB*, pointed at *A* and *B*, to which is attached a thick conical shell *S*, of brass, with flanges; a sliding weight *C* can be moved along the axis. Without the slider *C*, the centre of gravity of the top is nearly at the point *B*, and thus

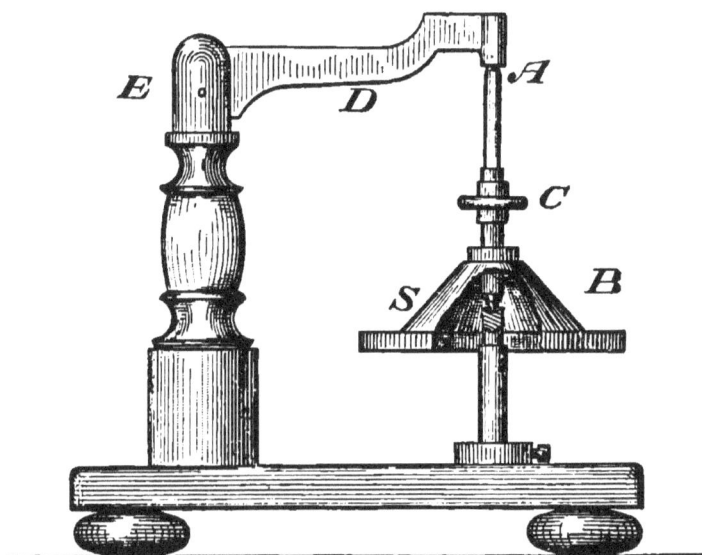

Fig. 51.

by moving *C* up and down, it can be made to fall either above or below the point *B*, or to coincide with it.

The top is spun by holding it in the position shown in the figure between an arm *ADE* (movable about a hinge at *E*) and a fixed upright with a small cup, roughened on the inside, in which the point *B* rests.

A string is wound about the axis, and the arm *ADE* being held lightly in position, the top is spun by pulling the string; and the arm being then removed, it remains spinning about the point *B* and exhibits the motions of precession indicated by the theory.

It may be noticed here that if the centre of gravity be exactly at the point B, and the top be accurately made, its axis will become a *permanent axis*, and no motion of precession will be seen, the top while spinning preserving the position initially given to it.

67. The motion of the top after it has been set spinning and placed on the plane, may be completely determined explicitly from the initial conditions and the two relations (a) and (b) just obtained :

$$\left\{ \begin{array}{l} A\left(\dfrac{d\theta}{dt}\right)^2 + A \sin^2\theta\left(\dfrac{d\psi}{dt}\right)^2 = 2\, Mgh(\cos\theta_0 - \cos\theta), \\[2ex] \qquad\qquad A \sin^2\theta\dfrac{d\psi}{dt} = Cn(\cos\theta_0 - \cos\theta). \end{array} \right.$$

These give $\dfrac{d\theta}{dt}$, $\dfrac{d\psi}{dt}$, and then the position and the motion of the top at every instant are known.

As we have seen, $\dfrac{d\psi}{dt}$ depends for its sign on n and the position of the centre of gravity; it also changes with θ; and, on eliminating $\dfrac{d\psi}{dt}$, we get

$$A \sin\theta\dfrac{d\theta}{dt} = \sqrt{\cos\theta_0 - \cos\theta}\,\sqrt{2Mgh\cdot A \sin^2\theta - C^2n^2(\cos\theta_0 - \cos\theta)};$$

and $\dfrac{d\theta}{dt}$ will also change in value, and will have minimum values (0) when $\theta = \theta_0$ and $\theta = \theta_1$, θ_1 being a root of the quadratic

$$2\,A\,Mgh\,\sin^2\theta - C^2n^2(\cos\theta_0 - \cos\theta) = 0.$$

The top will then, as it is first placed on the plane, tend to drop down, and $\dfrac{d\theta}{dt}$ will go on increasing until, having passed some maximum value, it reaches its minimum value when $\theta=\theta_1$.   Meanwhile $\frac{d\psi}{dt}$ has also been going through peri-odic changes, being a maximum when $\theta=\theta_1$.

The top then oscillates between the positions θ_0 and θ_1, and at the same time is carried about a vertical axis with a pre-cessional motion (not constant) $\frac{d\psi}{dt}$.

To an observer placed above the top and watching the pro-jection of its centre of gravity on the horizontal plane, that point would describe the curve indicated in Fig. 52, lying between two circles whose radii are $h\sin\theta_0$ and $h\sin\theta_1$.

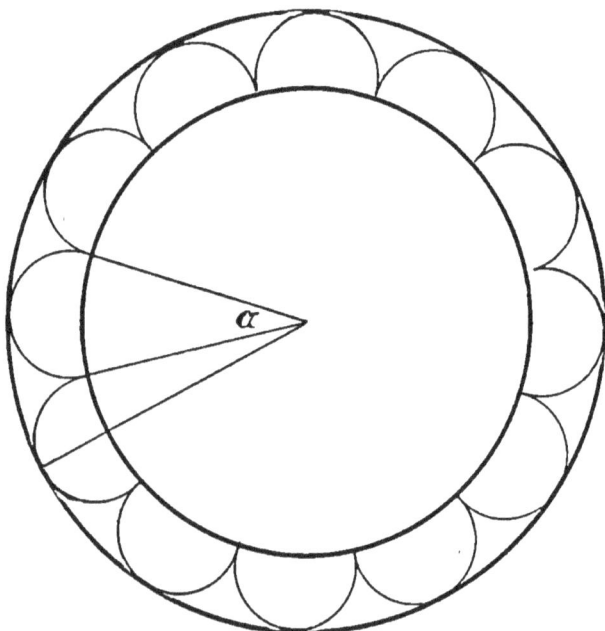

Fig. 52.

The curve described will not necessarily be closed; that will depend on α being an integral part of 2π. It is evident also, from the fact that maximum and minimum values exist at the cusps and the outer points, that the curve described touches one circle and cuts the other at right angles.

The maximum value of $\frac{d\psi}{dt}$ may be found by putting $\frac{d\theta}{dt}=0$ in the equations on page 118, which will give

$$A \sin^2 \theta \left(\frac{d\psi}{dt}\right)^2 = 2\, Mgh(\cos \theta_0 - \cos \theta),$$

$$A \sin^2 \theta \frac{d\psi}{dt} = Cn(\cos \theta_0 - \cos \theta).$$

$$\therefore \frac{d\psi}{dt} = \frac{2\, Mgh}{Cn} = \frac{2\, Wh}{Cn},$$

W being the weight of the top.

When $\theta = \theta_0$, it can be seen that both $\frac{d\theta}{dt}$ and $\frac{d\psi}{dt}$ vanish identically.

68. *Top spinning with Great Velocity on a Rough Horizontal Plane.*

In most cases the top is spun with a very great velocity, and then placed on the plane. By taking the value of $\frac{d\theta}{dt}$ already found,

$$A \sin \theta \frac{d\theta}{dt} = \sqrt{\cos \theta_0 - \cos \theta} \sqrt{2\, MghA \sin^2 \theta - C^2 n^2(\cos \theta_0 - \cos \theta)},$$

it will be seen that if n become very great, $\cos \theta_0 - \cos \theta$ must become very small in order that the expression under the radical may remain positive, hence the axis of the top, instead of performing large oscillations, will depart but little from θ_0, its initial position, and $\frac{d\psi}{dt}$ will approach a constant value, and the motion will therefore become steady. The time of a small oscillation may be found in the following way :

Let $\theta = \theta_0 + u$, u being small.

$$\therefore \frac{\cos\theta_0 - \cos\theta}{\sin\theta} = u$$

approximately, and the foregoing relation for $\dfrac{d\theta}{dt}$ becomes

$$A\frac{d\theta}{dt} = \sqrt{\frac{\cos\theta_0 - \cos\theta}{\sin\theta}}\sqrt{2\,MghA\sin\theta - C^2n^2\frac{\cos\theta_0 - \cos\theta}{\sin\theta}}.$$

$$\therefore A\frac{d\theta}{dt} = \sqrt{2\,MghA\sin\theta_0\,u - C^2n^2u^2}.$$

$$\therefore \frac{A}{Cn}\cdot\frac{d\theta}{dt} = \sqrt{2\,au - u^2},$$

where
$$a = \frac{Mgh\,A\sin\theta_0}{C^2n^2}.$$

But
$$\frac{d\theta}{dt} = \frac{du}{dt}.$$

$$\therefore \frac{A}{Cn}\cdot\frac{du}{\sqrt{2\,au - u^2}} = dt.$$

$$\therefore t = \frac{A}{Cn}\,\text{vers}^{-1}\left(\frac{u}{a}\right).$$

$$\therefore u = a\,\text{vers}\,\frac{Cn}{A}t,$$

and
$$\theta = \theta_0 + a\left(1 - \cos\frac{Cn}{A}t\right).$$

This is a periodic function which repeats values of θ every time t is increased by

$$\frac{2\pi}{\dfrac{Cn}{A}},$$

and therefore the time of a complete small oscillation is

$$\frac{2\pi A}{Cn}.$$

Also, $\dfrac{d\psi}{dt} = \dfrac{Cn}{A} \cdot \dfrac{\cos\theta_0 - \cos\theta}{\sin^2\theta} = \dfrac{Cn}{A} \cdot \dfrac{u}{\sin\theta_0}$

$$= \dfrac{Cn}{A\sin\theta_0} a\left(1 - \cos\dfrac{Cn}{A}t\right).$$

$$\therefore \dfrac{d\psi}{dt} = \dfrac{Cn}{A\sin\theta_0} \cdot \dfrac{Mgh\,A\sin\theta_0}{C^2n^2}\left(1 - \cos\dfrac{Cn}{A}t\right)$$

$$= \dfrac{Mgh}{Cn}\left(1 - \cos\dfrac{Cn}{A}t\right).$$

$$\therefore \psi = \dfrac{Mgh}{Cn}t - \dfrac{Mgh}{C^2n^2}A\sin\dfrac{Cn}{A}t,$$

and consist of two terms, one increasing uniformly with the time, the other very small, and a periodic function of the time.

If n be extremely large, we have, approximately,

$$\psi = \dfrac{Mgh}{Cn}t,$$

and the precession is then nearly constant and equal to

$$\dfrac{Wh}{Cn},$$

W being the weight of the top.

69. If, then, a top be spun with very great velocity and placed on a rough horizontal plane, inclined at an angle to the vertical, it will make small oscillations in time $\dfrac{2\pi A}{Cn}$, and at the same time will revolve about a vertical axis with an angular velocity very nearly equal to $\dfrac{Wh}{Cn}$. In the ordinary case, the oscillations will be so rapid at first as to be barely visible to the eye; as the speed diminishes, owing to resistance of air and friction at the apex, they become more noticeable; until finally,

when the top is "*dying*," n becomes comparable with the other quantities, the oscillations become wider, and the formulas of Art. 67 apply.

70. *Top spinning on a Smooth Horizontal Plane.*

Let a top (Fig. 53) be spun and placed in any manner on a smooth horizontal plane, and let its position after any time

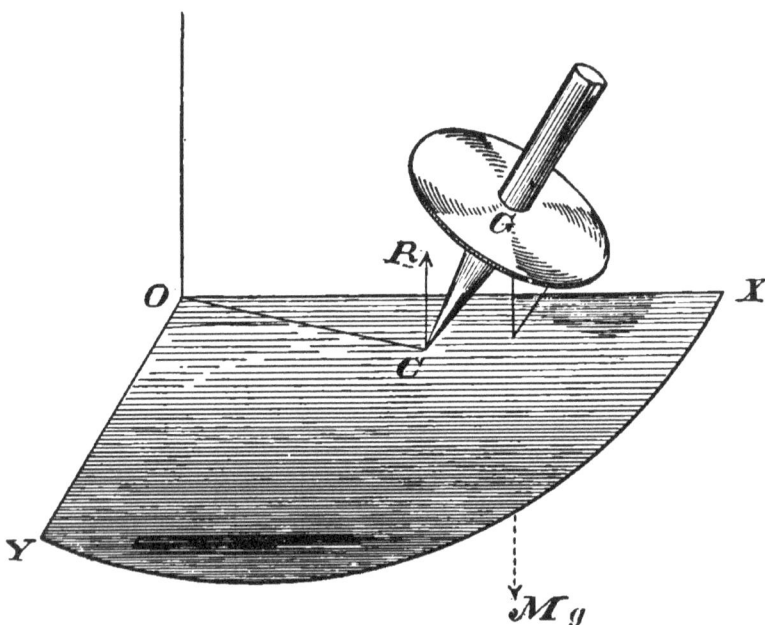

Fig. 53.

t has elapsed be that shown in the figure. It is acted upon only by the reaction R of the plane and its weight Mg acting at G, the centre of gravity; and if ξ, η, ζ be the coördinates of G, the equations of motion of translation are

$$\Sigma m \frac{d^2x}{dt^2} = M \frac{d^2\xi}{dt^2} = 0,$$

$$\Sigma m \frac{d^2y}{dt^2} = M \frac{d^2\eta}{dt^2} = 0,$$

$$\Sigma m \frac{d^2z}{dt^2} = M \frac{d^2\zeta}{dt^2} = R - Mg.$$

From these it is seen that $\frac{d\xi}{dt}$ = constant = initial value; $\frac{d\eta}{dt}$ = constant = initial value; and if therefore any horizontal motion be imparted initially to the centre of gravity, it will preserve that velocity at every instant thereafter.

And, since $\zeta = h \cos \theta$, CG being equal to h, and θ being the inclination of the axis of the top to the vertical, the third relation becomes

$$M \frac{d^2(h \cos \theta)}{dt^2} = R - Mg.$$

$$\therefore R = M \left\{ g + \frac{d^2(h \cos \theta)}{dt^2} \right\}.$$

The equations of motion of the top about the centre of gravity considered as a fixed point are

$$\text{(1)} \quad A \frac{d\omega_1}{dt} + (C - A)\omega_2\omega_3 = Rh \sin \theta \sin \phi,$$

$$\text{(2)} \quad A \frac{d\omega_2}{dt} + (A - C)\omega_1\omega_3 = Rh \sin \theta \cos \phi,$$

$$\text{(3)} \qquad\qquad\qquad \omega_3 = n ;$$

and we have also the relations

$$\text{(4)} \quad \frac{d\theta}{dt} = \omega_1 \sin \phi + \omega_2 \cos \phi,$$

$$\text{(5)} \quad \frac{d\psi}{dt} \sin \theta = \omega_2 \sin \phi - \omega_1 \cos \phi,$$

$$\text{(6)} \quad \frac{d\phi}{dt} + \frac{d\psi}{dt} \cos \theta = \omega_3 = n.$$

Thus it will be seen that, considering the centre of gravity as a fixed point, these equations are similar to those previously obtained in the case of a top spinning on a rough plane; the only difference being that for Mg in those relations we have R in these.

The solution is therefore similar to that given in Arts. 64 and 65.

We have
$$R = Mg + M \frac{d^2(h \cos \theta)}{dt^2}$$

$$= M \left\{ g - h \sin \theta \frac{d^2\theta}{dt^2} - h \cos \theta \left(\frac{d\theta}{dt}\right)^2 \right\}.$$

And multiplying (1) by ω_1 and (2) by ω_2, we have

$$A\omega_1 \frac{d\omega_1}{dt} + A\omega_2 \frac{d\omega_2}{dt} = Rh \sin \theta \{ \omega_1 \sin \phi + \omega_2 \cos \phi \}$$

$$= Rh \sin \theta \frac{d\theta}{dt}.$$

$$\therefore A(\omega_1^2 + \omega_2^2) = 2 \int Rh \sin \theta \frac{d\theta}{dt} dt$$

$$= 2 \int \left\{ Mgh \sin \theta - Mh^2 \sin^2 \theta \frac{d^2\theta}{dt^2} - Mh^2 \sin \theta \cos \theta \left(\frac{d\theta}{dt}\right)^2 \right\} d\theta.$$

$$\therefore A(\omega_1^2 + \omega_2^2) = 2 Mgh(\cos \theta_0 - \cos \theta) - Mh^2 \sin^2 \theta \left(\frac{d\theta}{dt}\right)^2.$$

But
$$\omega_1^2 + \omega_2^2 = \left(\frac{d\theta}{dt}\right)^2 + \left(\frac{d\psi}{dt}\right)^2 \sin^2 \theta.$$

$$\therefore A\left(\frac{d\theta}{dt}\right)^2 + A \sin^2 \theta \left(\frac{d\psi}{dt}\right)^2 + Mh^2 \sin^2 \theta \left(\frac{d\theta}{dt}\right)^2$$

$$= 2 Mgh(\cos \theta_0 - \cos \theta),$$

and the other relation will be as before:

$$A \sin^2 \theta \frac{d\psi}{dt} = Cn(\cos \theta_0 - \cos \theta).$$

These two relations give the solution of the problem.

And it is evident that, independent of its motion of translation in a horizontal direction, the centre of gravity can only move up and down with an oscillatory motion while the apex describes on the plane the fluted curve already obtained in the case of a rough plane (Fig. 52), the values of θ_0 and θ_1 being as before those which make $\dfrac{d\theta}{dt}$ a minimum.

If $\omega_3 = n$ be very great, the discussion is the same as before, and it can easily be seen that the apex of the top will describe a simple circle (approximately) on the plane, and the motion will be steady, the time of a small oscillation and the period of precession being obtained as formerly.

71. All the previous results obtained theoretically in the case of motion of a top on a smooth or rough plane can be verified experimentally by having a number of tops made similar to that shown in Fig. 54.

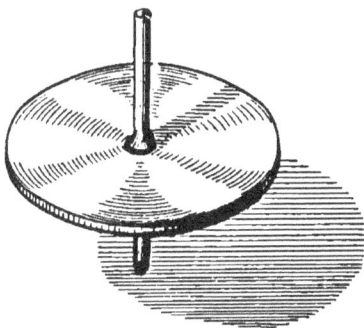

Fig. 54.

A circular plate of brass, a quarter of an inch in thickness, and from three to five inches in diameter, has a steel axis through the centre. The centre of gravity of the top may be from one to two inches from the apex on which it spins, and the point may have varying degrees of sharpness.

Everything should be symmetrical and made true, so that $A = B$.

The top is most readily set spinning by using a two-pronged handle with openings through which the axis may pass : a cord put through a hole in the axis and wound about it, is pulled rapidly, and the top drops with a high speed from the handle. A little practice enables one to spin the top and let it drop on a smooth or rough surface at any required inclination.

The following problem may also be examined by using several of these tops of various sizes, and with points of varying degrees of sharpness :

A common top, when spun and placed on a rough horizontal plane, at an angle to the vertical, gradually assumes an upright position. Explain this.

This is the case of the ordinary peg top of the schoolboy, which is usually made of a cone of wood through which passes a steel axis ending in a sharp point ; when spun upon a rather rough surface, it gradually becomes upright and 'sleeps.'

It will be found, after a little experimenting, that this apparently paradoxical rising of the top to a vertical position against the force of gravity depends on two things :

1. *The degree of sharpness of the apex on which the top spins.*

2. *The position of the centre of gravity.*

If the point be very sharp so that the top in spinning is not able to form a small conical bed for itself and thereby be acted on by a couple arising from friction at a considerable distance from the point, it cannot possibly become erect.

When, however, the point is rather blunt, and the centre of gravity not too high, the top will slowly rise up under the action of the friction (which tends to diminish the angle of inclination), and 'sleep.'

The equations of motion are similar to those obtained in Art. 64, with the additional relations introduced by friction.

The solution of the equations shows that the top rises to the vertical, on the supposition that the point of the top is a portion

of a spherical surface and that friction is thus enabled to act in
the proper manner.

A complete analytical solution of the problem is given in
Jellett's *Theory of Friction*, Chap. VIII., where the top is supposed to be a symmetrical pear-shaped cone with a spherical
surface as the apex upon which it spins.

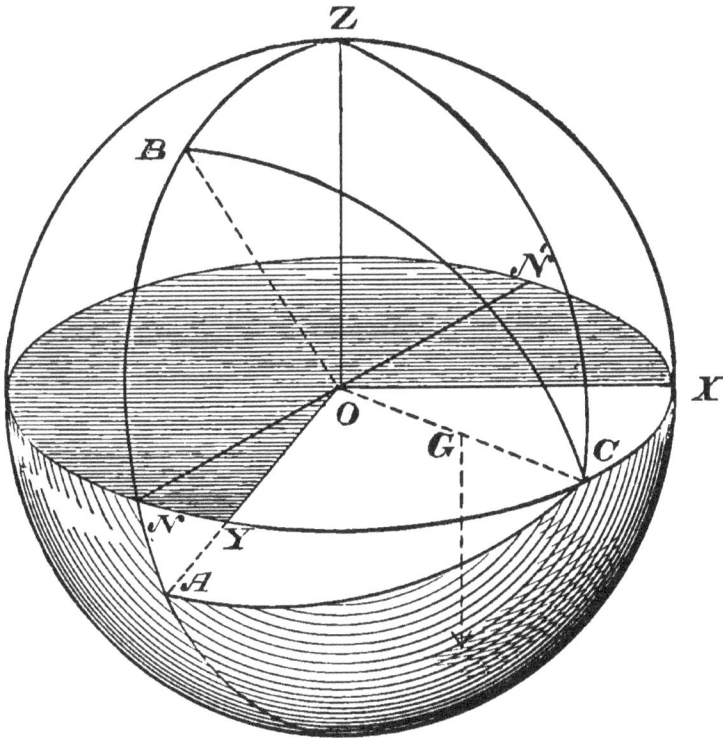

Fig. 55.

72. *The Gyroscope moving in a Horizontal Plane about a
Fixed Point.*

If a gyroscope be put in rapid motion and placed so that the
prolongation of the axis of rotation can rest on a fixed point of
support, and if, at the same time, an initial angular velocity

about the point of support be given bodily to the gyroscope (in the proper direction) in a horizontal plane, it will revolve about a vertical axis, and the apparently paradoxical motion is presented of a body whose centre of gravity moves in a horizontal plane although its point of support is at quite a distance from the vertical through the centre of gravity.

In Fig. 55 the gyroscope is supposed to be set rotating and started in a horizontal plane with its centre of gravity at the point G, the weight acting vertically downwards in the direction indicated by the arrow. It is supported only at the point O, and, if rotating rapidly enough, will keep on moving uniformly in this horizontal plane in a direction hereafter determined.

Its position at any time is given by the position of its *principal axes* at O: these are OA, OB, OC.

It is evident that $\theta = \frac{\pi}{2}$ and that C moves along XNN', NON' being the line of nodes, and the angle $BON = \phi$.

At each instant the gyroscope tends bodily to turn about NON' under the action of gravity, and the value of this turning couple is mgh, m being the mass of the gyroscope and $OG = h$.

Resolving this couple mgh into two, we get

$$mgh \cos \phi \text{ about } OB,$$

and $$mgh \sin \phi \text{ about } OA.$$

Then Euler's equations become:

$$\begin{cases} A\dfrac{d\omega_1}{dt} + (C - A)\omega_2\omega_3 = mgh \sin \phi, \\[2mm] A\dfrac{d\omega_2}{dt} - (C - A)\omega_1\omega_3 = mgh \cos \phi, \\[2mm] C\dfrac{d\omega_3}{dt} = 0, \end{cases}$$

from which it is seen that

$$\omega_3 = \text{constant} = n.$$

K

Also, we have

$$\left\{\begin{array}{l} \dfrac{d\theta}{dt}=0=\omega_1 \sin\phi+\omega_2 \cos\phi, \\[2mm] \dfrac{d\psi}{dt}=\omega_2 \sin\phi-\omega_1 \cos\phi, \\[2mm] \dfrac{d\phi}{dt}=n, \end{array}\right.$$

since $\theta=\dfrac{\pi}{2}$, and therefore $\cos\theta=0$.

From the preceding relations we have :

$$\left.\begin{array}{l} \omega_1 \sin\phi+\omega_2 \cos\phi=0, \\[2mm] \omega_2 \sin\phi-\omega_1 \cos\phi=\dfrac{d\psi}{dt}. \end{array}\right\}$$

$\therefore$ squaring and adding

$$\omega_1{}^2+\omega_2{}^2=\left(\dfrac{d\psi}{dt}\right)^2.$$

But since

$$A\dfrac{d\omega_1}{dt}+(C-A)\omega_2\omega_3=mgh \sin\phi,$$

$$A\dfrac{d\omega_1}{dt}-(C-A)\omega_1\omega_3=mgh \cos\phi.$$

Therefore, multiplying the former by ω_1, and the latter by ω_2, and adding, we get

$$A\omega_1\dfrac{d\omega_1}{dt}+A\omega_2\dfrac{d\omega_2}{dt}=mgh(\omega_1 \sin\phi+\omega_2 \cos\phi)=0.$$

$$\therefore A(\omega_1{}^2+\omega_2{}^2)=\text{constant}.$$

$$\therefore \omega_1{}^2+\omega_2{}^2=\text{its initial value,}=a^2 \text{ say}.$$

Then
$$\dfrac{d\psi}{dt}=a, \text{ and } \dfrac{d\phi}{dt}=n.$$

$$\therefore \ \phi = nt, \ \psi = \alpha t,$$

since both may be taken zero when t is zero.

$$\left.\begin{aligned} \therefore \ \omega_1 &= -\alpha \cos nt, \\ \omega_2 &= \alpha \sin nt, \\ \omega_3 &= n. \end{aligned}\right\}$$

And, substituting these values in the relation for the first couple, we get

$$A(\alpha n \sin nt) + (C - A)n(\alpha \sin nt) = mgh \sin nt.$$

$$\therefore \ Cn\alpha \sin nt = mgh \sin nt.$$

$$\therefore \ Cn\alpha = mgh,$$

and
$$\alpha = \frac{d\psi}{dt} = \frac{mgh}{Cn} = \frac{Wh}{Cn},$$

W being the weight of the top, and n being the initial velocity of rotation. Hence the axis OC moves around in a horizontal plane with uniform velocity $\dfrac{Wh}{Cn}$, and the direction of revolution is indicated by the sign of n or ω_3; that is, *to an observer looking down in the direction ZO, the gyroscope will revolve bodily in the same direction as the gyroscope rotates about its axis when viewed by an observer at C.*

It is important to observe that the necessary condition for the motion of the gyroscope bodily about OZ is that it receives an initial angular velocity, so that

$$\omega_1{}^2 + \omega_2{}^2 = \left(\frac{d\psi}{dt}\right)^2 = \text{some finite quantity.}$$

If this initial velocity be not given to it, it will act in the same way as a top, tending to drop down and oscillate as it moves around the vertical.

RIGID DYNAMICS.

Usually n is very great, so that α is small, and the precessional motion is slow.

For a complete discussion of the experiments which can be performed with the *Gyroscope* see Chap. X.

73. *To find the Pressure on the Fixed Point in the Case of the Gyroscope.*

As an illustration of the use of the equations of Art. 63, we may find the pressure on the point about which the gyroscope revolves.

In this case we shall have, calling the mass of the gyroscope S to avoid confusion,

$$
S\left\{\omega_1 \cdot \overline{B+C-A}\left(\frac{y\omega_2}{C}+\frac{z\omega_3}{B}\right)-(\omega_2^2+\omega_3^2)x\right\}
$$
$$
=P\cos\lambda+\Sigma m X-S\left(\frac{M}{B}z-\frac{N}{C}y\right),
$$
$$
S\left\{\omega_2 \cdot \overline{C+A-B}\left(\frac{z\omega_3}{A}+\frac{x\omega_1}{C}\right)-(\omega_3^2+\omega_1^2)y\right\}
$$
$$
=P\cos\mu+\Sigma m Y-S\left(\frac{N}{C}x-\frac{L}{A}z\right),
$$
$$
S\left\{\omega_3 \cdot \overline{A+B-C}\left(\frac{x\omega_1}{B}+\frac{y\omega_2}{A}\right)-(\omega_1^2+\omega_2^2)z\right\}
$$
$$
=P\cos\nu+\Sigma m Z-S\left(\frac{L}{A}y-\frac{M}{B}x\right),
$$

which become, since $A=B$, and (x, y, z) are $(0, 0, h)$,

$$
S\left\{\omega_1\cdot C\cdot\frac{hn}{A}\right\}=P\cos\lambda+Sg\cos\phi-S\left\{\frac{Sg\cos\phi}{A}\cdot h^2\right\},
$$
$$
S\left\{\omega_2\cdot C\cdot\frac{hn}{A}\right\}=P\cos\mu+Sg\sin\phi+S\left\{\frac{Sg\sin\phi}{A}\cdot h^2\right\},
$$
$$
S\{-(\omega_1^2+\omega_2^2)h\}=P\cos\nu,
$$

the last of which can be obtained from elementary considerations.

$$\therefore \quad \begin{cases} P\cos\lambda = m\left\{\dfrac{mg\cos\phi}{A}\cdot h^2 - g\cos\phi + \omega_1 C\dfrac{hn}{A}\right\}, \\[2mm] P\cos\mu = m\left\{-\dfrac{mg\sin\phi}{A}\cdot h^2 - g\sin\phi + \omega_2 C\dfrac{hn}{A}\right\}, \\[2mm] P\cos\nu = m\left\{-(\omega_1{}^2 + \omega_2{}^2)h\right\}, \end{cases}$$

the mass being denoted by m.

These relations taken in conjunction with

$$\begin{cases} \omega_1\sin\phi + \omega_2\cos\phi = 0, \\[2mm] \omega_2\sin\phi - \omega_1\cos\phi = \dfrac{d\psi}{dt} = \dfrac{Wh}{Cn}, \\[2mm] \dfrac{d\phi}{dt} = n, \end{cases}$$

give, on squaring and adding, the value of P in terms of known quantities.

Similar equations may be obtained in the case of the top spinning on a smooth horizontal plane.

CHAPTER VII.

MOTION ABOUT A FIXED POINT.

Impulsive Forces.

74. In forming the general equations of motion for finite forces, we had two sets of relations of which the types are

$$\Sigma m \frac{d^2x}{dt^2} = \frac{d}{dt} \Sigma m \frac{dx}{dt} = \Sigma m X + P \cos \lambda,$$

and $\quad \Sigma m \left\{ y \frac{d^2z}{dt^2} - z \frac{d^2y}{dt^2} \right\} = \frac{d}{dt} \Sigma m \left\{ y \frac{dz}{dt} - z \frac{dy}{dt} \right\} = L,$

and, remembering the definition of an impulse, we get our impulsive equations from these by integrating with respect to t, from o to τ, some small value of the time.

That is, instead of a continuous change we have an abrupt change of velocity and of moment of momentum taking place during an exceedingly small time τ.

Hence, for impulses X, Y, Z, we get the equations

$$\Sigma m \left\{ \left(\frac{dx}{dt} \right)' - \left(\frac{dx}{dt} \right) \right\} = \Sigma m \left\{ (z\omega_y' - y\omega_z') - (z\omega_y - y\omega_z) \right\}$$

$$= (\omega_y' - \omega_y) \Sigma m z - (\omega_z' - \omega_z) \Sigma m y$$

$$= M\bar{z} \cdot (\omega_y' - \omega_y) - M\bar{y}(\omega_z' - \omega_z)$$

$$= \Sigma X + P \cos \lambda,$$

with two similar relations for Y and Z. These determine the impulse P.

For the couples we have

$$\left[\Sigma m\left\{y\frac{dz}{dt}-z\frac{dy}{dt}\right\}\right]'-\left[\Sigma m\left\{y\frac{dz}{dt}-z\frac{dy}{dt}\right\}\right]=L.$$

But

$$\Sigma m\left(y\frac{dz}{dt}-z\frac{dy}{dt}\right)=\Sigma m\{y(y\omega_z-x\omega_y)-z(x\omega_z-z\omega_x)\}$$

$$=\omega_z\Sigma m\overline{y^2+z^2}-\omega_y\Sigma mxy-\omega_z\Sigma mxz$$

$$=A\omega_z-F\omega_y-E\omega_z.$$

∴ we get, for the three couples,

$$\begin{cases} A(\omega_z'-\omega_z)-F(\omega_y'-\omega_y)-E(\omega_z'-\omega_z)=L, \\ B(\omega_y'-\omega_y)-D(\omega_z'-\omega_z)-F(\omega_z'-\omega_z)=M, \\ C(\omega_z'-\omega_z)-E(\omega_z'-\omega_z)-D(\omega_y'-\omega_y)=N, \end{cases}$$

ω_z, ω_y, ω_z being the angular velocities about axes fixed in space at time t, and these being suddenly changed by the impulsive actions to ω_z', ω_y', ω_z'.

75. Taking the foregoing expressions for the impulsive couples, we can simplify them by choosing principal axes, which make D, E, F vanish; if, at the same time, the body starts from rest, ω_z, ω_y, ω_z are zero, and the equations become

$$\begin{cases} A\omega'_z=L, \\ B\omega'_y=M, \\ C\omega'_z=N. \end{cases}$$

The equations of the instantaneous axis are

$$\frac{x}{\omega'_z}=\frac{y}{\omega'_y}=\frac{z}{\omega'_z},$$

or

$$\frac{x}{\dfrac{L}{A}}=\frac{y}{\dfrac{M}{B}}=\frac{z}{\dfrac{N}{C}},$$

or

$$\frac{Ax}{L}=\frac{By}{M}=\frac{Cz}{N}.$$

The plane of the impulsive couple is

$$Lx + My + Nz = 0,$$

and therefore the instantaneous axis (that is, the line about which the body will begin to rotate under the action of the impulse) is the line conjugate to the plane

$$Lx + My + Nz = 0$$

with regard to the ellipsoid

$$Ax^2 + By^2 + Cz^2 = c.$$

The equations of the instantaneous axis are

$$\frac{Ax}{L} = \frac{By}{M} = \frac{Cz}{N},$$

and the equations of the axis of the impulsive couple are

$$\frac{x}{L} = \frac{y}{M} = \frac{z}{N}.$$

Hence it will be seen, by comparing these two sets of relations, that if a body fixed at a point be struck, it will not begin to rotate about the axis of the impulsive couple induced by the blow, unless $A = B = C$, or unless the plane of the impulsive couple be a principal plane or parallel to a principal plane. For the two sets cannot reduce to a single set unless $A = B = C$, or unless two of the quantities, x, y, z, vanish, (which means that the axis of the couple is one of the principal axes).

It will be seen from the preceding investigation that, if a rigid body be free to turn about a fixed point, the problem of determining the change produced in the motion of the body by the action of a given impulse, is equivalent to determining the change in its motion when the body is acted on by a given impulsive couple. This equivalence also appears from the following considerations. The impulse may be resolved into an equal and parallel impulse acting at the fixed point and an impulsive couple. The impulse acting at the fixed point will have no influence on the motion of the body, and therefore only the couple need be considered. Resolving the latter with respect to the coördinate axes we obtain the equations on page 135.

Illustrative Examples.

1. A cube is fixed at its centre of inertia, and struck along an edge.

In this simple case it is evident, without forming the equations of motion, that, since the momental ellipsoid is a sphere, $A = B = C$, and the cube begins to rotate about the axis of the impulsive couple.

Thus, in Fig. 56, the cube is fixed at O, its centre of inertia, and on being struck by a blow Q, begins to rotate about the axis of the impulsive couple AOB.

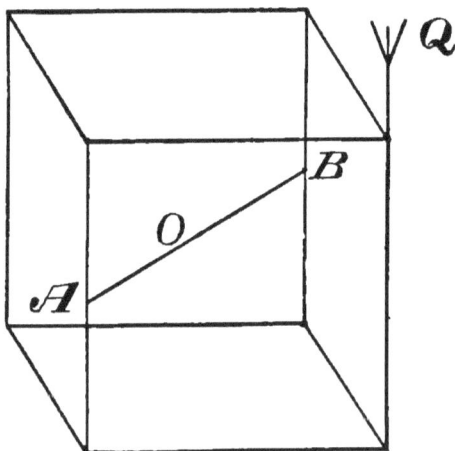

Fig. 56.

2. A homogeneous solid right circular cylinder is rotating with given angular velocity about its centre of inertia, which is fixed ; the cylinder receives a blow of given intensity in a direc-

tion perpendicular to the plane in which its axis moves. Determine the subsequent motion.

3. A lamina in the form of a semi-ellipse bounded by the axis minor is movable about the centre as a fixed point, and falls from the position in which its plane is horizontal; determine the impulse which must be applied at the centre of inertia, when the lamina is vertical, in order to reduce it to rest.

If this impulse be applied perpendicularly to the lamina, at the extremity of an ordinate, through the centre of inertia, instead of being applied at the centre of inertia itself, show that the lamina will begin to revolve about the major axis.

4. A triangular plate (right angled) fixed at its centre of inertia and struck at the right angle perpendicularly to the plate.

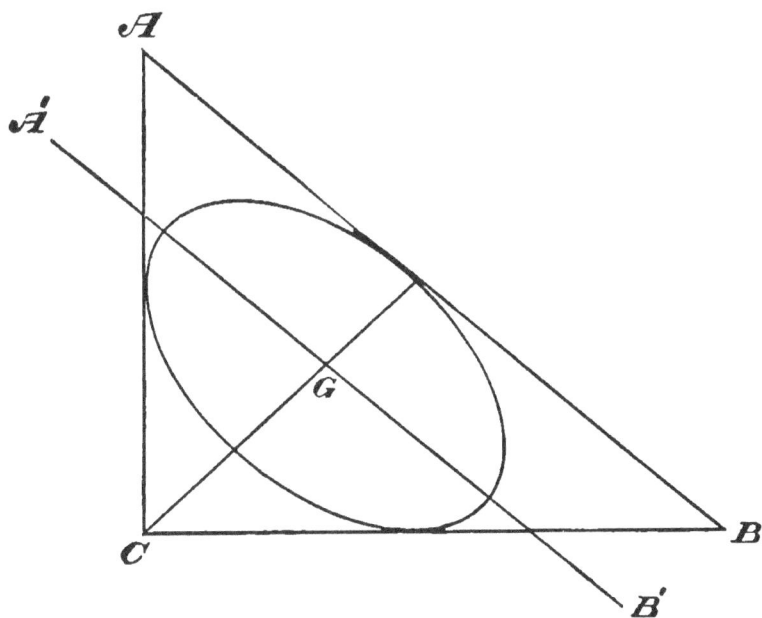

Fig. 57.

In Fig. 57 let G be the centre of inertia of the triangle, and C the point where the blow is struck at right angles to the plane of the paper. Then if we construct the momental ellipse at G, it touches the three sides at their middle points. The impulsive couple in this case contains the line CG in its plane ; but since AB is a tangent to the ellipse, $A'GB'$ is the diametral line conjugate to CG. The triangle therefore commences to rotate about $A'GB'$, which is drawn parallel to the hypothenuse.

5. A solid ellipsoid fixed at its centre is struck normally at a point p, q, r.

If l, m, n, be the direction cosines of the line of the blow whose magnitude is Q, and if the equation of the ellipsoid be

$$\frac{x^2}{a^2}+\frac{y^2}{b^2}+\frac{z^2}{c^2}=1,$$

then the equations of the instantaneous axis will be

$$\frac{Ax}{L}=\frac{By}{M}=\frac{Cz}{N},$$

or

$$\frac{\dfrac{M}{5}(b^2+c^2)x}{Q(qn-rm)}=\cdots=\cdots,$$

or

$$\frac{(b^2+c^2)x}{qn-rm}=\frac{(c^2+a^2)y}{rl-pn}=\frac{(a^2+b^2)z}{pm-ql},$$

and since the blow is normal to the ellipsoid at p, q, r,

$$\frac{l}{\dfrac{du}{dx}}=\frac{m}{\dfrac{du}{dy}}=\frac{n}{\dfrac{du}{dz}},$$

or

$$\frac{l}{\dfrac{p}{a^2}}=\frac{m}{\dfrac{q}{b^2}}=\frac{n}{\dfrac{r}{c^2}}.$$

Therefore the equations of the instantaneous axis will be

$$\frac{p}{a^2}\cdot\frac{b^2+c^2}{b^2-c^2}x=\frac{q}{b^2}\cdot\frac{c^2+a^2}{c^2-a^2}y=\frac{r}{c^2}\cdot\frac{a^2+b^2}{a^2-b^2}\cdot z.$$

CHAPTER VIII.

MOTION ABOUT A FIXED POINT. NO FORCES ACTING.

76. *Heavy Body fixed at its Centre of Gravity.*

The simplest case of motion under no forces which ordinarily presents itself is that of a body acted on by gravity and fixed in such a manner that it can only rotate about its centre of gravity considered as a fixed point.

Here we have

$$A \frac{d\omega_1}{dt} - (B - C)\omega_2\omega_3 = 0,$$

$$B \frac{d\omega_2}{dt} - (C - A)\omega_3\omega_1 = 0,$$

$$C \frac{d\omega_3}{dt} - (A - B)\omega_1\omega_2 = 0.$$

And, multiplying these three equations by ω_1, ω_2, ω_3, respectively, and adding, we get

$$A\omega_1 \frac{d\omega_1}{dt} + B\omega_2 \frac{d\omega_2}{dt} + C\omega_3 \frac{d\omega_3}{dt} = 0.$$

$$\therefore A\omega_1{}^2 + B\omega_2{}^2 + C\omega_3{}^2 = \text{a constant}$$

$$= k^2. \tag{1}$$

Similarly, multiplying the three equations by $A\omega_1$, $B\omega_2$, $C\omega_3$, respectively, adding, and integrating, we get

$$A^2\omega_1{}^2 + B^2\omega_2{}^2 + C^2\omega_3{}^2 = \text{a constant}$$

$$= h^2. \tag{2}$$

140

(1) states that the kinetic energy is constant, as might be expected, since no forces act ; this can be seen by taking

$$\tfrac{1}{2}\Sigma mv^2 = \tfrac{1}{2}\Sigma m\left\{\left(\frac{dx}{dt}\right)^2 + \left(\frac{dy}{dt}\right)^2 + \left(\frac{dz}{dt}\right)^2\right\}$$

$$= \tfrac{1}{2}\Sigma m\{(z\omega_2 - y\omega_3)^2 + \cdots + \cdots\}^2$$

$$= \tfrac{1}{2}\{A\omega_1^2 + B\omega_2^2 + C\omega_3^2\},$$

since the products of inertia vanish.

(2) is another way of expressing the constancy of the moment of momentum.

For (moment of momentum)^2

$$= h_1^2 + h_2^2 + h_3^2$$

$$= h^2,$$

where

$$h_1 = \Sigma m\left\{y\frac{dz}{dt} - z\frac{dy}{dt}\right\} = A\omega_1,$$

$$h_2 = \Sigma m\left\{z\frac{dx}{dt} - x\frac{dz}{dt}\right\} = B\omega_2,$$

$$h_3 = \Sigma m\left\{x\frac{dy}{dt} - y\frac{dx}{dt}\right\} = C\omega_3.$$

77. Now, since h_1, h_2, h_3 are constant at all times, the plane $h_1x + h_2y + h_3z = 0$, or $A\omega_1 x + B\omega_2 y + C\omega_3 z = 0$ is an *Invariable Plane* fixed in space ; the line

$$\frac{x}{A\omega_1} = \frac{y}{B\omega_2} = \frac{z}{C\omega_3}$$

is perpendicular to this plane, and is an *Invariable Axis*.

The *instantaneous axis* is given by

$$\frac{x}{\omega_1} = \frac{y}{\omega_2} = \frac{z}{\omega_3} = \frac{r}{\omega}.$$

78. If we now construct the momental ellipsoid at the fixed point O, as in Fig. 58, OA, OB, OC being the principal axes,

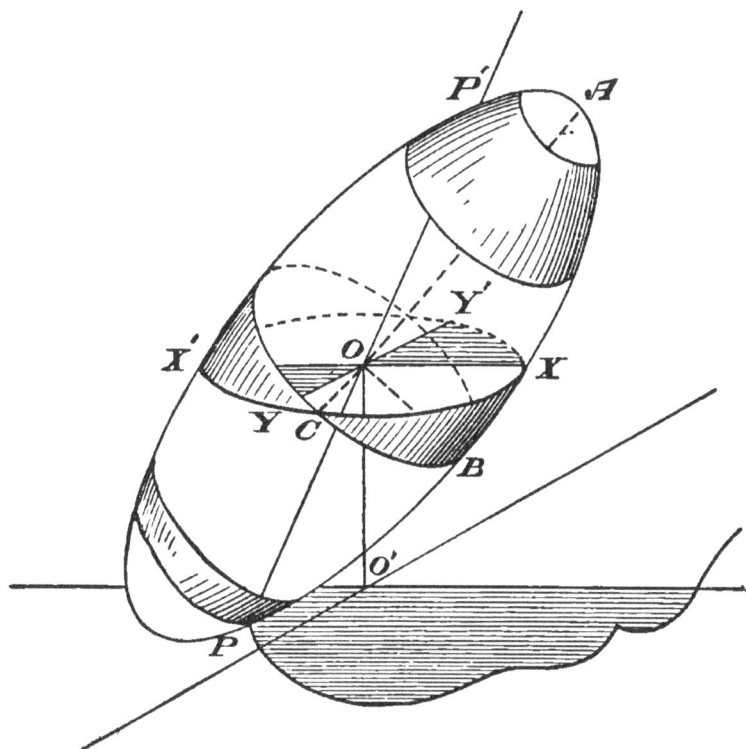

Fig. 58.

and POP' the instantaneous axis at any time t, the equation of the ellipsoid will be

$$Ax^2 + By^2 + Cz^2 = c^2,$$

and those of the instantaneous axis

$$\frac{x}{\omega_1} = \frac{y}{\omega_2} = \frac{z}{\omega_3} = \frac{r}{\omega}.$$

Now, x, y, z being any point on this line, let it represent the point P; then at P we have

$$\frac{x}{\omega_1} = \frac{y}{\omega_2} = \frac{z}{\omega_3} = \frac{r}{\omega}$$

$$= \frac{\sqrt{Ax^2 + By^2 + Cz^2}}{\sqrt{A\omega_1^2 + B\omega_2^2 + C\omega_3^2}} = \frac{c}{k}.$$

$$\therefore \ \omega = \frac{k}{c} \cdot r,$$

and

$$x = \omega_1 \cdot \frac{c}{k},$$

$$y = \omega_2 \cdot \frac{c}{k},$$

$$z = \omega_3 \cdot \frac{c}{k}.$$

Therefore the angular velocity at any instant is proportional to the radius vector of the ellipsoid.

Moreover, taking the tangent plane at P to the ellipsoid, its equation is

$$(\xi - x)\frac{du}{dx} + (\eta - y)\frac{du}{dy} + (\zeta - z)\frac{du}{dz} = 0,$$

where

$$x = \omega_1 \frac{c}{k}, \ y = \omega_2 \frac{c}{k}, \ z = \omega_3 \frac{c}{k},$$

which becomes

$$\left(\xi - \omega_1 \frac{c}{k}\right)\frac{du}{dx} + \cdots + \cdots = 0,$$

or

$$\left(\xi - \omega_1 \cdot \frac{c}{k}\right) 2 A \frac{c}{k} \omega_1 + \cdots + \cdots = 0,$$

or

$$A\omega_1 \cdot \xi + B\omega_2 \cdot \eta + C\omega_3 \cdot \zeta = kc.$$

And, if we construct the plane

$$A\omega_1 x + B\omega_2 y + C\omega_3 z = 0$$

and represent it by $XYX'Y'$, this is the *invariable plane;* and we see that the tangent plane to the momental ellipsoid at the point where the instantaneous axis cuts the ellipsoid is always parallel to this invariable plane.

Hence, the motion of the body fixed at O, and under the action of no forces, is completely represented *by the rolling of the momental ellipsoid on a plane fixed in space* and parallel to the invariable plane, and at a distance from it equal to OO'.

79. The ellipsoid in rolling on the fixed plane traces out a curve on that plane, and also one on its own surface.

The curve traced out on the surface of the ellipsoid is called the *Polhode*, and its equation is found by taking the condition that the perpendicular from the centre of the ellipsoid on a tangent plane at x, y, z is constant, and combining it with the equation of the ellipsoid itself.

The equation of the *Polhode* is, therefore,

$$Ax^2 + By^2 + Cz^2 = c^2,$$
$$A^2x^2 + B^2y^2 + C^2z^2 = c'^2.$$

The curve traced out on the plane is called the *Herpolhode*, and its equation is found from the relation

$$O'P^2 = \rho^2 = OP^2 - OO'^2 = r^2 - p^2,$$

and will vary with r, and therefore with ω and with p.

It is apparent that any one of the central ellipsoids might be chosen instead of the momental ellipsoid, and the motion of the body exhibited in a similar manner by the changes in motion of the ellipsoid chosen.

Innumerable problems may be constructed from the preceding representation ; but they are all dependent on properties of the ellipsoid, and are not problems in Dynamics.

MOTION OF A FREE BODY.

80. We have already seen, in discussing *D'Alembert's Principle*, that the general equations of motion of any body are

$$\Sigma m\left(X-\frac{d^2x}{dt^2}\right)=0,$$

$$\Sigma m\left(Y-\frac{d^2y}{dt^2}\right)=0,$$

$$\Sigma m\left(Z-\frac{d^2z}{dt^2}\right)=0,$$

and

$$\Sigma m\left\{y\left(Z-\frac{d^2z}{dt^2}\right)-z\left(Y-\frac{d^2y}{dt^2}\right)\right\}=0,$$

$$\Sigma m\left\{z\left(X-\frac{d^2x}{dt^2}\right)-x\left(Z-\frac{d^2z}{dt^2}\right)\right\}=0,$$

$$\Sigma m\left\{x\left(Y-\frac{d^2y}{dt^2}\right)-y\left(X-\frac{d^2x}{dt^2}\right)\right\}=0.$$

If M be the whole mass, $\bar{x}$, $\bar{y}$, $\bar{z}$ the coördinates of the centre of inertia at time t, and x', y', z' the place of m relatively to a system of axes originating at the centre of inertia and parallel to the original set of axes, then the equations of motion become

$$M\frac{d^2\bar{x}}{dt^2}=\Sigma mX,$$

$$M\frac{d^2\bar{y}}{dt^2}=\Sigma mY,$$

$$M\frac{d^2\bar{z}}{dt^2}=\Sigma mZ,$$

and

$$\left\{ \Sigma m \left\{ y' \left(Z - \frac{d^2 z'}{dt^2} \right) - z' \left(Y - \frac{d^2 y'}{dt^2} \right) \right\} = 0, \right.$$

$$\Sigma m \left\{ z' \left(X - \frac{d^2 x'}{dt^2} \right) - x' \left(Z - \frac{d^2 z'}{dt^2} \right) \right\} = 0,$$

$$\left. \Sigma m \left\{ x' \left(Y - \frac{d^2 y'}{dt^2} \right) - y' \left(X - \frac{d^2 x'}{dt^2} \right) \right\} = 0, \right.$$

which latter can be transformed in the ordinary way so as to determine the angular velocities.

These equations theoretically give a complete solution of the problem.

But the most important case of free motion of a body, and the only one which admits of simple solution, is that in which

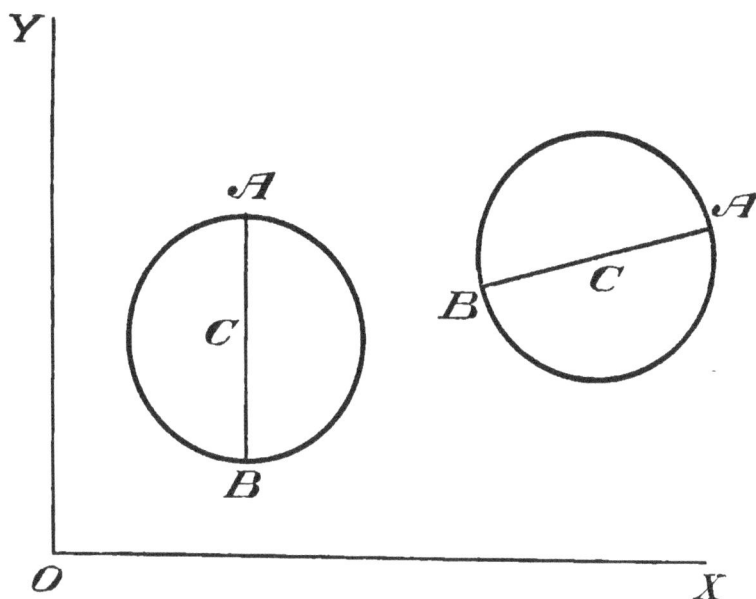

Fig. 59.

the particles of the body move in parallel planes. Here it is evident that we need only consider the motion of one particular plane of particles, and that containing the centre of inertia is chosen, and the position of the body at any time determined in the following way.

Let the plane in which the centre of inertia moves be represented by the plane of the paper, the same section of the body being represented at any two times as in Fig. 59.

Let the body be referred to fixed axes OX, OY, and let ACB be any line in the body passing through the centre of inertia C, and in its initial position let this line be parallel to OY, as shown. Then, after any time t, the body has reached its second position, and it is evident from elementary geometry that the body can get from its first position to the second by translation of the centre of inertia C, and by rotation about C through an angle θ, equal to that which ACB in its second position makes with the axis OY, or with a parallel line fixed in space.

For translation of the centre of inertia, we have, by D'Alembert's principle,

$$\left\{ \begin{aligned} \Sigma m \frac{d^2x}{dt^2} &= \Sigma mX = M \frac{d^2\bar{x}}{dt^2}, \\ \Sigma m \frac{d^2y}{dt^2} &= \Sigma m Y = M \frac{d^2\bar{y}}{dt^2}. \end{aligned} \right.$$

And for rotation about the centre of inertia considered as a fixed point, we get

$$\Sigma m \left\{ x \frac{d^2y}{dt^2} - y \frac{d^2x}{dt^2} \right\} = \Sigma m r^2 \frac{d^2\theta}{dt^2} = N.$$

Therefore, at any time, the motion of the body will be fully known when we know

1. The initial conditions, so that θ is known.

2. The coördinates of the centre of inertia with reference to some axes fixed in space ; this gives $\dfrac{d^2\bar{x}}{dt^2}$, $\dfrac{d^2\bar{y}}{dt^2}$.

3. Mk^2 about the axis of rotation through the centre of inertia.

4. Geometrical relations between $\bar{x}$, $\bar{y}$, θ

In cases of constraint where bodies roll or slide on others, geometrical relations are easily found, and the unknown reactions eliminated by taking moments.

Illustrative Examples.

1. A heavy sphere rolling down a perfectly rough inclined plane.

In this problem gravity, by the aid of friction and the reaction of the plane, produces both the translation of the centre of inertia and the rotation.

Let OX, OY (Fig. 60) be the axes of the coördinates fixed in space, the sphere starting to roll from O. Then at any time t,

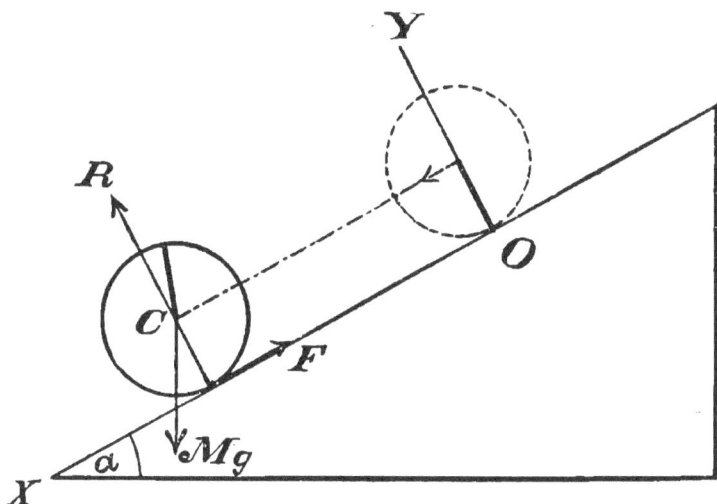

Fig. 60

the position of the sphere, is given by $\bar{x}$, $\bar{y}$, the coördinates of C, the centre of inertia, and the angle θ through which the sphere has rolled; that is, through which it has rotated about C, considered as a fixed point.

The initial conditions, combined with the geometrical conditions for perfect rolling, give

$$\bar{x}=a\theta, \ \bar{y}=a. \tag{1}$$

For the translation of the centre of inertia, we have

$$\left\{ \begin{array}{l} M\dfrac{d^2\bar{x}}{dt^2}+F-Mg\sin\alpha=0, \qquad\qquad (2) \\[3mm] M\dfrac{d^2\bar{y}}{dt^2}+Mg\cos\alpha-R=0. \qquad\quad (3) \end{array}\right.$$

The rotation about C is given by

$$Mk^2\frac{d^2\theta}{dt^2}=Fa. \qquad\qquad (4)$$

These four relations give a complete solution of the problem, for we have

$$\frac{d^2\bar{y}}{dt^2}=0, \quad \frac{d^2\bar{x}}{dt^2}=a\frac{d^2\theta}{dt^2};$$

and, therefore, from (2) and (4),

$$M(k^2+a^2)\frac{d^2\theta}{dt^2}=Mga\sin\alpha, \qquad\qquad (5)$$

from which it is seen that

$$\frac{d^2\bar{x}}{dt^2}=\tfrac{5}{7}g\sin\alpha,$$

and

$$\bar{x}=\tfrac{5}{14}g\sin\alpha\cdot t^2;$$

also,

$$R=Mg\cos\alpha,$$

$$F=\tfrac{2}{7}Mg\sin\alpha.$$

These results give the space passed over in time t, and show that five-sevenths of gravity is used in translation, while two-sevenths is used in turning the sphere about the centre of inertia.

The relation (5) may also be obtained at once by forming the *equation of energy*. For the sphere has fallen through a distance $\bar{x}\sin\alpha$, and therefore the work done by gravity is

$$Mg\bar{x} \sin \alpha,$$

or $$Mga\theta \sin \alpha,$$

which must be equal to the kinetic energy at time t, and therefore

$$\tfrac{1}{2} M(v^2 + k^2\omega^2) = Mga\theta \sin \alpha.$$

$$\therefore \ \tfrac{1}{2} M(a^2 + k^2)\left(\frac{d\theta}{dt}\right)^2 = Mga\theta \sin \alpha,$$

which gives, on differentiation,

$$M(a^2 + k^2)\frac{d^2\theta}{dt^2} = Mga \sin \alpha,$$

as before.

2. If a heavy circular cylinder rolls down a perfectly rough inclined plane, one-third of gravity is used in turning and two-thirds in translation.

3. A very thin spherical shell surrounds a sphere, both being perfectly smooth and consequently no friction between them, and the system rolls down a rough inclined plane.

In this case, if we neglect the mass of the outer shell, the inner sphere acts just as if it slid down the plane, because, since there is no friction between it and the shell, as the shell rolls it slips around, and therefore the equation of motion is

$$M\frac{d^2x}{dt^2} = Mg \sin \alpha,$$

M being the mass of the sphere, which is so large that the mass of the outer shell is negligible in comparison.

If, however, the shell and sphere were united, the system would roll down, and then the equation of motion would be

$$M\frac{d^2x}{dt^2} = M \tfrac{5}{7} g \sin \alpha.$$

And the times occupied in rolling a given distance in the two cases would be to one another as $\sqrt{5} : \sqrt{7}$.

In the case of a cylinder surrounded by a cylindrical shell, gravity would be diminished to two-thirds of its value, and the times occupied in rolling a given distance would be as $\sqrt{2} : \sqrt{3}$ under similar circumstances.

4. To determine whether a sphere is hollow or solid by rolling it down a rough plane. This could be done by observing the space passed over in a given time, and by calculating the moments of inertia and forming the equations of motion (1) on the supposition of a solid body; (2) on the supposition of a shell of radii a, b.

5. A homogeneous heavy sphere rolls down within a rough spherical bowl; it is required to determine the motion.

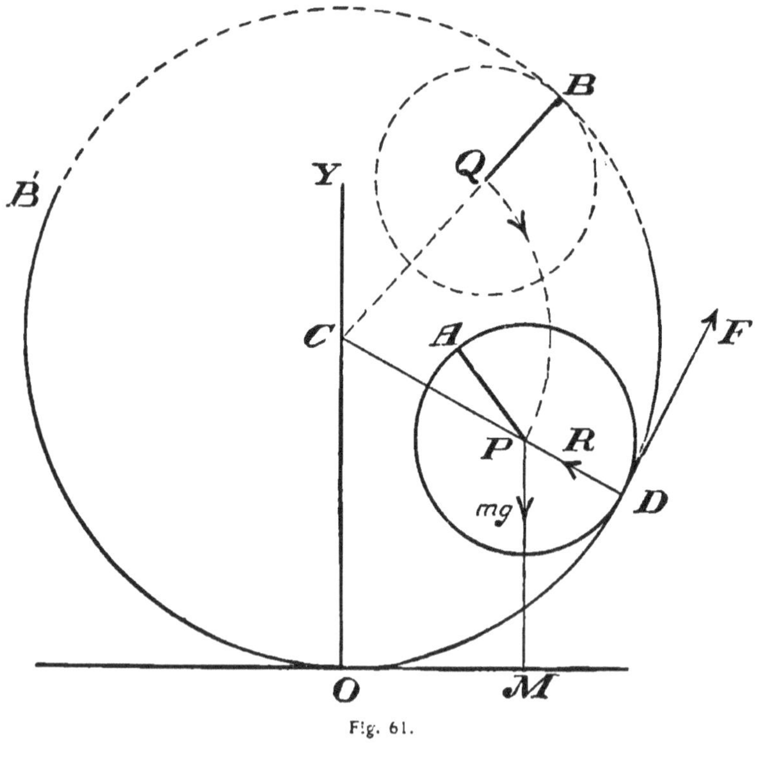

Fig. 61.

Let the radius of the spherical bowl (Fig. 61) be b, and of the sphere, a; and let the sphere start with AP coincident with BQ. Then, at time t, the circumstances are as shown in the figure.

Let $\qquad \omega =$ angular velocity about P,

$$OCP = \phi, \qquad\qquad OM = \bar{x},$$

$$DPA = \theta, \qquad\qquad PM = \bar{y},$$

$$BCO = \alpha,$$

then will $\qquad \bar{x} = (b-a)\sin\phi,$

and $\qquad \bar{y} = b - (b-a)\cos\phi.$

And the equations of motion are

$$M\frac{d^2\bar{x}}{dt^2} = -R\sin\phi + F\cos\phi, \qquad (1)$$

$$M\frac{d^2\bar{y}}{dt^2} = R\cos\phi + F\sin\phi - mg, \qquad (2)$$

F being the friction, and R the reaction at the point D, acting in the directions indicated by the arrows.

Moreover, $\qquad \omega = \dfrac{d(MPA)}{dt} = \dfrac{d(\phi+\theta)}{dt} = \dfrac{d\phi}{dt} + \dfrac{d\theta}{dt},$

MPA being the exterior angle at P, and

$$a\theta = b(\alpha - \phi),$$

$$\therefore \ \omega = \frac{a-b}{a}\cdot\frac{d\phi}{dt}. \qquad (3)$$

Along with the foregoing relation we have, also, taking moments about P,

$$MK^2\frac{d\omega}{dt} = F\cdot a. \qquad (4)$$

It is then easy to find R and F by taking the values of $\bar{x}$ and $\bar{y}$, and differentiating twice and substituting in (1) and (2).

It will be found on reduction, that

$$R = \frac{mg}{7}(17 \cos \phi - 10 \cos \alpha),$$

and $$(b-a)\left(\frac{d\phi}{dt}\right)^2 = \tfrac{10}{7}g(\cos \phi - \cos \alpha).$$

From this latter expression, by differentiation, we get

$$\frac{d^2\phi}{dt^2} = -\frac{5}{7} \cdot \frac{g}{b-a}\sin \phi,$$

and if ϕ becomes small, this gives the *time of a small oscillation* of a sphere within a spherical bowl. For

$$\frac{d^2\phi}{dt^2} + \frac{5}{7} \cdot \frac{g}{b-a} \cdot \phi = 0,$$

represents a motion of oscillation of which the periodic time is

$$2\pi\sqrt{\frac{7(b-a)}{5g}}.$$

It may also be noticed that the pressure on the bowl vanishes when $\cos \phi = \tfrac{10}{17} \cos \alpha$.

If BOB' were completed and the sphere supposed to rotate about C with angular velocity sufficient to keep the smaller sphere at the top, the pressure against the outer sphere and the conditions of equilibrium can at once be found from the relations already obtained, which also furnish a solution to the following instructive problem :

6. A perfectly rough ball is placed within a hollow cylindrical garden roller at the lowest point, and the roller is then drawn along a level walk with a uniform velocity V. Show that the ball will roll quite round the interior of the roller if V^2 be $>$ $\tfrac{27}{7} g(b-a)$, a being the radius of the ball, and g of the roller.

7. A uniform straight rod slips down in a vertical plane
between two smooth planes, one horizontal, the other vertical;
find the motion.

Let OX, OY be the horizontal and vertical planes, and let
the rod starting from its upper position when $t = 0$ assume the
position AB at time t, as in Fig. 62.

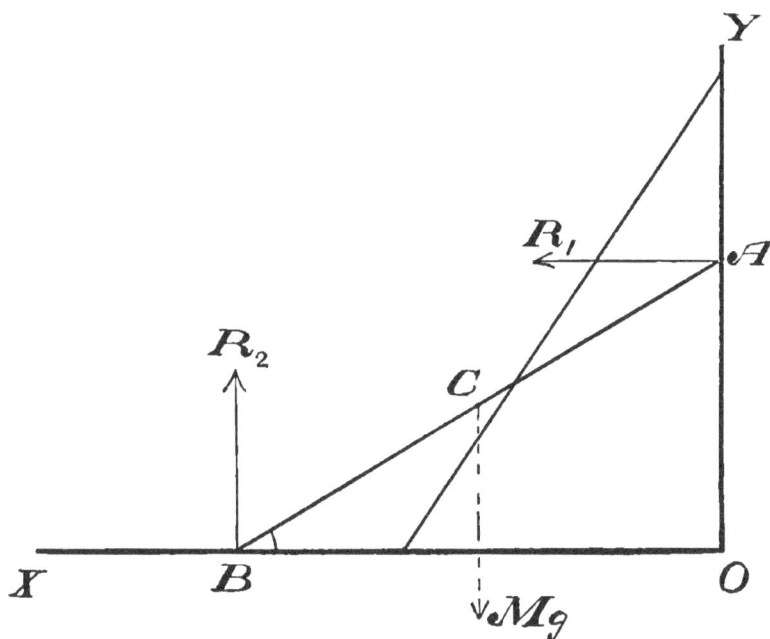

Fig. 62.

Then we have two reactions at the points A and B, and the
weight Mg acting at the centre of gravity, C.

So that if $\bar{x}, \bar{y}$ be the coördinates of C, and θ the angle of
inclination of the rod AB to the horizontal, we get

$$\begin{cases} \Sigma m \dfrac{d^2 x}{dt^2} = M \dfrac{d^2 \bar{x}}{dt^2} = R_1, \\[2ex] \Sigma m \dfrac{d^2 y}{dt^2} = M \dfrac{d^2 \bar{y}}{dt^2} = R_2 - Mg, \end{cases}$$

and
$$\bar{x} = a \cos \theta, \\ \bar{y} = a \sin \theta,$$

where $2\,a$ is the length of the rod.

Also, taking moments about the centre of gravity, we would have

$$R_2 \cdot a \cos \theta + Mk^2 \cdot \frac{d^2\theta}{dt^2} = R_1 \cdot a \sin \theta,$$

and we may suppose that θ is initially equal to α.

These four relations give a complete solution of the problem.

It will be found that the rod leaves the vertical plane when $\sin \theta = \frac{2}{3} \sin \alpha$, and then the motion becomes changed, the rod moving with a constant horizontal velocity along the horizontal plane equal to $\sqrt{\dfrac{2\,ga\,(\sin \alpha)^3}{3}}$, until it finally drops and lies in the plane.

The problem may also be solved by aid of the *principle of energy*.

8. A circular disc capable of motion about a vertical axis through its centre perpendicular to its plane is set in motion with angular velocity Ω. A rough uniform sphere is gently placed on any point of the disc, not the centre; prove that the sphere will describe a circle on the disc, and that the disc will revolve with angular velocity $\dfrac{7\,Mk^2}{7\,Mk^2 + 2\,mr^2} \cdot \Omega$, where Mk^2 is the moment of inertia of the disc about its centre, m is the mass of the sphere, and r is the radius of the circle traced out.

9. A homogeneous sphere is placed at rest on a rough inclined plane, the coefficient of friction being μ; determine whether the sphere will slide or roll.

10. A homogeneous sphere is placed on a rough table, the coefficient of friction being μ, and a particle one-tenth of the mass of the sphere is attached to the extremity of a horizontal

diameter. Show that the sphere will begin to roll or slide according as $\mu >$ or $< \dfrac{11}{10\sqrt{37}}$. What will happen if $\mu = \dfrac{11}{10\sqrt{37}}$?

81. *Impulsive Actions. Motion of a Billiard Ball.*

The complex motions of a homogeneous sphere moving on a rough horizontal plane are well illustrated in the game of billiards, where an ivory sphere is struck by a cue and made to perform evolutions that seem to the unscientific little short of marvellous.

In the general case the course which the billiard ball takes depends on the initial circumstances, that is to say, on the way in which it is struck by the cue ; and the motion is made up of both sliding and rolling, so that the centre of the ball moves in a portion of a parabola until the sliding motion ceases, when it rolls on in a straight line. If struck so that the cue is in the same vertical plane with the centre of the sphere, then the motion is purely rectilinear ; which is also the case if the cue is held in a horizontal position.

It may also happen that if. the ball be struck by the cue at a certain oblique inclination to the table, its path, after sliding ceases, will be opposite to the horizontal direction of the stroke, and it will roll backwards.

For a complete solution of the problem, then, we should know the direction, intensity, and point of application of the blow struck by the cue, so that the velocity of translation of the centre of gravity is known, and the initial angular velocity.

82. In ordinary blows, the initial value of the rolling friction will be very small compared with the sliding friction, so that at the beginning the former may be neglected, and the equations of motion for sliding found in the following way.

Let the plane in which the centre of the ball moves be the plane of xy, so that $(x, y, -a)$ are the coördinates of the point of contact at time t. Let F be the value of the sliding friction, and β the angle it makes with the axis of x.

Then evidently the pressure on the table is equal to the weight of the ball, so that $R = Mg$ and $F = \mu R$.

The equations of motion of the centre of gravity are

$$M\frac{d^2x}{dt^2} = -F\cos\beta,$$
$$M\frac{d^2y}{dt^2} = -F\sin\beta,$$
$$M\frac{d^2z}{dt^2} = 0 = R - Mg.$$

For rotation about the centre of gravity we have

$$A\frac{d\omega_1}{dt} = -aF\sin\beta,$$
$$A\frac{d\omega_2}{dt} = aF\cos\beta,$$
$$A\frac{d\omega_3}{dt} = 0.$$

$$\therefore A(\omega_1 - \Omega_1) = aM\left(\frac{dy}{dt} - v_0\right),$$
$$A(\omega_2 - \Omega_2) = -aM\left(\frac{dx}{dt} - u_0\right),$$
$$\omega_3 = \Omega_3,$$

where u_0, v_0 are the axial components of the initial velocities of the centre of gravity, and Ω_1, Ω_2, Ω_3 are the initial angular velocities about axes through the centre of gravity.

The above give a complete solution of the motion during sliding which, however, in the case of an ordinary billiard ball, lasts but for a small fraction of a second.

83. At the instant the ball is struck by the cue the *impulsive equations* will evidently be formed as follows.

Let Q be the value of the blow struck by the cue, and α the angle the cue makes with the table; also, let F be the *impulsive value of friction* at the instant of striking, and β the angle which it makes with the axis of x.

Then, the axes being chosen as in the preceding problem, and the line and angular velocities being denoted as formerly by u_0, v_0, Ω_1, Ω_2, Ω_3, we have

$$\left\{ \begin{array}{l} Mu_0 = Q \cos \alpha - F \cos \beta, \\ Mv_0 = \qquad\quad - F \sin \beta, \end{array} \right.$$

$$\left\{ \begin{array}{l} A\Omega_1 = - Qh \sin \alpha - aF \sin \beta, \\ A\Omega_2 = \quad Qk \quad + aF \cos \beta, \\ A\Omega_3 = - Qh \cos \alpha, \end{array} \right.$$

where h is the horizontal distance from the centre of the ball to the vertical plane containing the line of blow, and k is the perpendicular on the line of blow from the point where h meets the vertical plane containing that line. And the impulse on the table must be equal to $Q \sin \alpha$. See, *Théorie mathématique des effets du jeu de billard, par G. Coriolis, Paris*, 1835.

84. *Impulsive Actions. Free Body. Illustrative Examples.*

1. A uniform rod is lying on a smooth horizontal table and is struck at one end in a direction perpendicular to its length. Determine the motion.

What if it be struck at the centre, or at the centre of percussion for a rotation-axis through one end of the rod?

2. Two uniform rods of equal length are freely hinged together and placed in a straight line on a smooth horizontal plane. The system is then struck at one end in a direction perpendicular to its length. Examine the motion initially and subsequently.

Here, the circumstances are a little more complicated than in the preceding problem, so that it is well to form the equations of motion of the two rods separately.

Let m be the mass of each rod, $2a$ the length, C, C' the centres of gravity, v, v' the velocities of translation of C, C', and ω, ω' the angular velocities.

Then if O, O' be the instantaneous centres so that $CO = x$ and $C'O' = x'$, we get

$$\left.\begin{array}{l} \omega x = v, \\ \omega' x' = v', \end{array}\right\}$$

and
$$(a - x)\omega = (a + x')\omega'.$$

And if Q be the blow, and R the reaction at the free hinge, the equations of motion of the two rods are

$$\left.\begin{array}{l} mv = Q + R, \\ \dfrac{ma\omega}{3} = Q - R, \end{array}\right\}$$

$$\left.\begin{array}{l} mv' = R, \\ \dfrac{ma\omega'}{3} = R, \end{array}\right\}$$

from which it will be found that

$$\omega = 2\omega',$$

and the initial velocity of the end struck is four times that of the other end.

3. Three uniform and equal rods AB, BC, CD are arranged as three sides of a square having free hinges at B and C; the end A is struck in the plane of the rods and at right angles to AB by a blow Q. Determine the motion, and show that the initial velocity of A is nineteen times that of D.

This is solved in the same way as the former problem by considering each portion separately. Thus, if R be the reaction of B, we have

$$
\left.\begin{array}{c}
mv = Q + R, \\[2mm]
\dfrac{ma\omega}{3} = Q - R,
\end{array}\right\}
$$

and $\omega x = v.$

Also, if R' be the reaction at C,

$$
R - R' = m(a - x)\omega = m(a + x')\omega',
$$

since the displacements of B and C are equal and in the same direction.

And for CD, $\left.\begin{array}{c} mv' = R', \\[2mm] \dfrac{ma\omega'}{3} = R'. \end{array}\right\}$

4. If in the preceding problem BC be a thin string whose mass is negligible, show that the initial velocity of A will be seven times that of D.

This is evident, for $R = R'$.

5. Two equal uniform rods AB, BC, freely jointed at B, are placed on a smooth horizontal table at right angles to one another, and a blow is applied at A perpendicular to AB; prove that the initial velocities of A, C are as 8 to 1.

6. Four equal uniform rods AB, BC, CD, DE, freely jointed at B, C, D, are laid on a horizontal table in the form of a square, and a blow is applied at A at right angles to AB from the inside of the square; prove that the initial velocity of A is 79 times that of E.

7. Three equal inelastic rods of length a, freely hinged together, are placed in a straight line on a smooth horizontal plane, and the two outer ends are set in motion about the ends

of the middle rod with equal but opposite angular velocities (ω) ; show' that after impact the triangle formed by the three will move on with a velocity $\frac{1}{3} a\omega$.

8. Four equal rods freely jointed together so as to form a square are moving with given velocity in the direction of a diagonal of the square, on a smooth horizontal plane. If one end of this diagonal impinge directly on an inelastic obstacle, find the time in which the rods will be in one straight line.

9. Four equal uniform rods each connected by a hinge at one extremity with the middle point of the rod next in order, initially form a square with produced sides, and are in motion with a given velocity in direction parallel to one of the rods. If an impulse be given at the free extremity of this rod, and the centre of inertia of the system be thereby reduced to rest, find the initial angular velocities of the four rods, and prove that these angular velocities remain unchanged during the subsequent motion.

10. A lamina in the form of an ellipse is rotating in its own plane with angular velocity ω about a focus. Suddenly this focus is freed and the other fixed. Find the velocity about the second focus.

M

CHAPTER X.

THE GYROSCOPE.

85. This instrument, to which reference has already been made in connection with motion about a fixed point, consists essentially of a wheel which is put in rotation within an outer ring: the latter being provided with knife edges and other arrangements whereby the whole mass may be experimented upon while the wheel is kept in motion.

A type of gyroscope, known as *Foucault's*, is shown in Fig. 63, and also more in detail in Figs. 65 and 66.

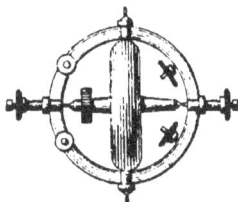

Fig. 63.

It is made of a disc, turned to offer the least resistance to the air, which can be made to rotate with great speed (from two hundred and fifty to five hundred times per second) about an axis through its centre of gravity.

This is done by means of the wheelwork motor (driven by hand) shown in Fig. 64, which is geared up at the top to the small toothed cog-wheel seen in Fig. 63, at the left-hand side of the disc, on the axis of the gyroscope, and within the outer ring.

162

The axis of rotation is of course movable in the outer ring, and this latter is provided with two knife edges which should be exactly in the prolongation of a line passing through the centre of gravity and perpendicular to the rotation axis.

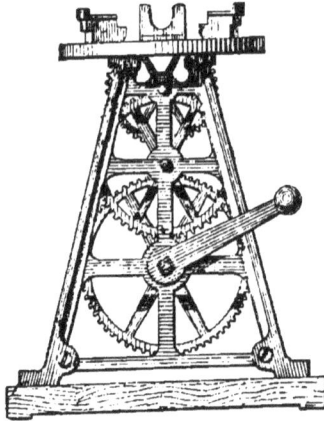

Fig. 64.

Four movable masses, two within the ring, and two outside, Fig. 65, are used to adjust the instrument in two perpendicular planes, so that the centre of gravity of the system will be in the line of the knife edges.

It is quite a difficult matter to perform this adjustment, which must be exact; since the slightest deviation of the position of the centre of gravity from this line destroys the value of the results obtained in the pendulum experiment.

The readiest way to adjust the gyroscope is to let it oscillate, under the action of gravity, about the knife edges, the centre of gravity being arranged at first to fall below the line of the knife edges (by properly altering the positions of the movable masses); and then, by slight variations of these positions, to bring the centre of gravity up until the oscillations about the knife-edge axis are made in from eight to ten seconds: the line of the knife edges is in that case infinitely close to the centre of gravity and the equilibrium nearly neutral.

86. *The Gyroscope moving in a Horizontal Plane about a Fixed Point.*

The gyroscope being adjusted, the experiment indicated by the theory of Art. 72 may easily be performed.

It is only necessary to place the instrument on top of the motor so that the wheels are properly geared, and to set the disc in rapid rotation, taking care that the bearings are carefully cleaned and oiled.

Then, placing it as shown in Fig. 65, so that a small pointed hook which is directly in the prolongation of the axis of rota-

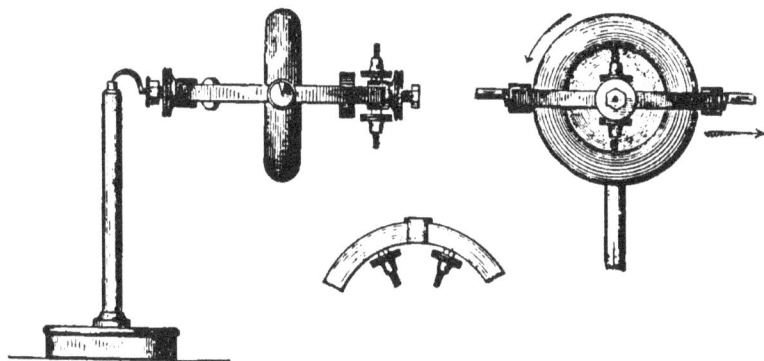

Fig. 65.   Fig. 66.

tion rests on a little agate cup at the top of an upright stand, the instrument is given a slight angular displacement bodily about a vertical axis passing through the point at which the hook rests, and it slowly moves about the vertical with an angular velocity equal to that found by the theory of Art. 72.

Moreover, the direction of motion is as shown in Fig. 66; that is, the gyroscope moves bodily about a vertical axis (when viewed from above) in the same direction as the disc rotates when viewed by an observer looking towards the fixed point about which the motion takes place.

Thus there is a perfect accord between theory and experiment, and the truth of the fundamental equations of motion is established.

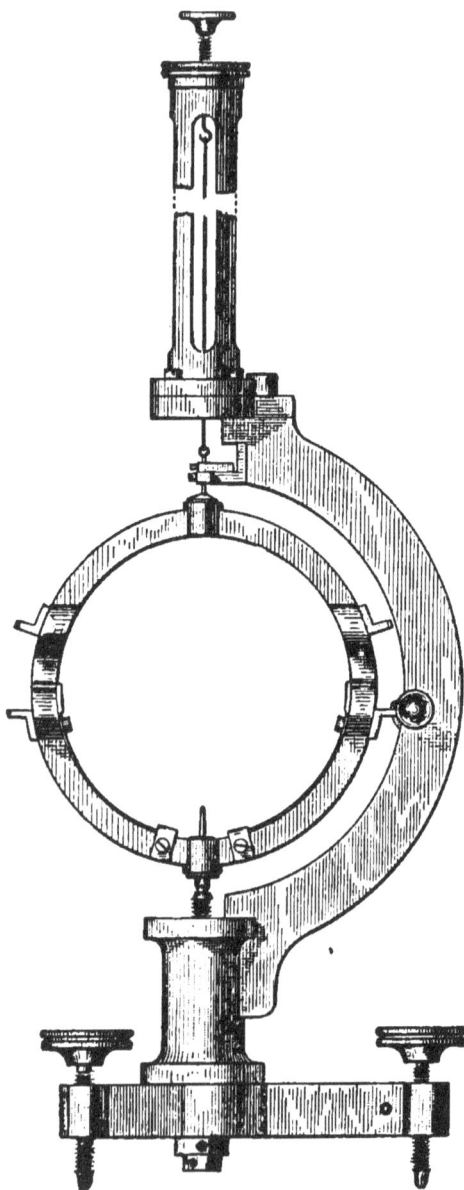

Fig. 67.

It may be observed also that if the gyroscope be given no initial impulse, but be merely let drop, it will act in the same manner as a top, and oscillate up and down while it keeps in motion about the vertical.

87. *To prove the Rotation of the Earth upon its Axis.*

This experiment depends on the permanency of the rotation axis in space.

A stand with pendulum is arranged as shown in Fig. 67.

There is a ring suspended by means of a fibre without torsion from a hook above, and the whole being carefully levelled so that the line of suspension is vertical, the gyroscope is put in rapid rotation and placed in the ring with the knife edges resting within beds provided for them : the ring, being then released by the small screw seen at the right, is quite free in space, and owing to the rapid rotation of the disc the axis of rotation is a *permanent axis* and remains fixed in space.

Hence, while the earth moves along, carrying with it the stand and observer, the gyroscope preserves its position in space for some time ; and if a long index be attached to it in prolongation of the rotation axis or parallel to it, this index will have an apparent motion from east to west, as the observer is carried along with the earth from west to east.

If the pendulum with the gyroscope were placed at the north pole, it is evident that the apparent motion of the index would be 360° in twenty-four hours.

At the equator there would be no apparent motion ; as although a permanent axis would still exist, the earth would simply carry the whole instrument bodily about the rotation axis of the earth.

Action in Any Latitude λ.

To find the angular velocity of the gyroscope in any latitude, let *PCF*, Fig. 68, be the axis of rotation of the earth.

And let the gyroscope be suspended at *A*, in the tangent plane, and preferably let the plane of rotation of the disc be in the geographical meridian plane.

Then the angular velocity of the earth about PCF is

$$\omega = 360° \text{ in twenty-four hours.}$$

And this, if resolved along CA, will produce a rotation about CA equal to $\omega \sin \lambda$, and this is the component which affects the gyroscope at A.

Since ω is against the hands of a watch, looking towards C from P, therefore $\omega \sin \lambda$, looking from A' towards A or C, will be against the hands of a watch, and therefore if Fig. 69

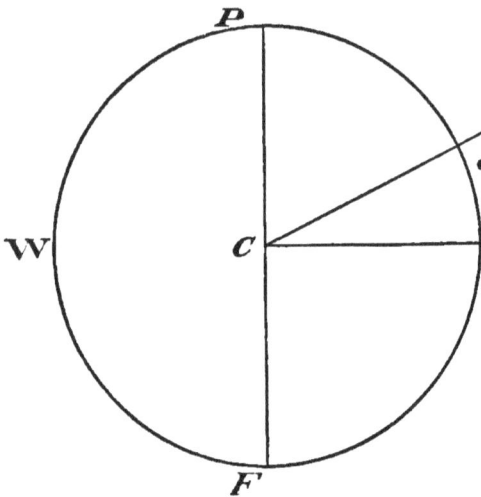

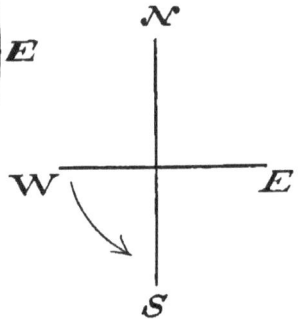

Fig. 68. Fig. 69.

represents the tangent plane at A, to an observer at A' above the gyroscope, the earth will move from west to east as indicated by the arrow, and the apparent motion of the index attached to the gyroscope will be as before from east to west.

88. It is evident also that the angular velocity being $\omega \sin \lambda$, if this be observed by noting the time and the angle passed over in that time, since ω is known to be 360° in twenty-four hours, we get a method for finding λ, the latitude of the place of experiment.

89. *Electrical Gyroscope.*

The defect of Foucault's gyroscope being that it does not keep up its motion long enough to give marked results in the pendulum experiment, an electrical gyroscope has been devised by *Mr. Hopkins*, who gives a description of his instrument in the *Scientific American* of July 6, 1878, and also in his recent text-book on Physics. His instrument is shown in Fig. 70.

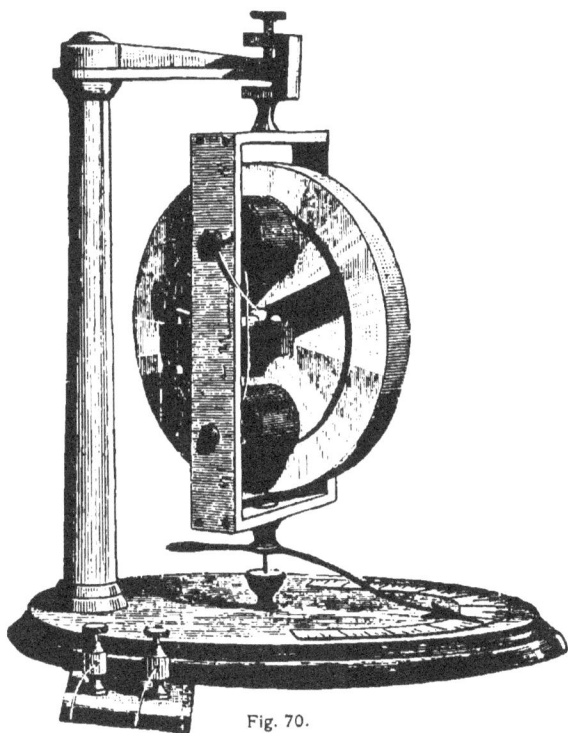

Fig. 70.

The rectangular frame which contains the wheel is supported by a fine and very hard steel point, which rests upon an agate step in the bottom of a small iron cup at the end of the arm that is supported by the standard. The wheel spindle turns on carefully made steel points, and upon it are placed two cams, one at each end, which operate the current-breaking springs.

The horizontal sides of the frame are of brass, and the vertical sides are iron. To the vertical sides are attached the cores of the electro-magnets. There are two helices and two cores on each side of the wheel, and the wheel has attached to it two armatures, one on each side, which are arranged at right angles to each other. The two magnets are oppositely arranged in respect of polarity, to render the instrument astatic.

An insulated stud projects from the middle of the lower end of the frame to receive an index that extends nearly to the periphery of the circular base piece and moves over a graduated semicircular scale. An iron point projects from the insulated stud into a mercury cup in the centre of the base piece, and is in electrical communication with the platinum pointed screws of the current breakers. The current-breaking springs are connected with the terminals of the magnet wires, and the magnets are in electrical communication with the wheel-supporting frame. One of the binding posts is connected by a wire with the mercury in the cup, and the other is connected with the standard. A drop of mercury is placed in the cup that contains the agate step to form an electrical connection between the iron cup and the pointed screw.

The current breaker is contrived to make and break the current at the proper instant, so that the full effect of the magnets is realized, and when the binding posts are connected with four or six Bunsen cells the wheel rotates at a high velocity.

The wheel will maintain its plane of rotation, and when it is brought into the plane of the meridian, the index will appear to move slowly over the scale in a direction contrary to the earth's rotation, but in reality the earth and the scale with it move from west to east, while the index remains nearly stationary.

90. *Fessel's Gyroscope.*

Another most useful and instructive form of gyroscope is that known as Fessel's, which is represented in Fig. 71.

" *Q* is a heavy fixed stand, the vertical shaft of which is a cylinder bored smoothly, in which works a vertical rod CC', as far as possible without friction, carrying at its upper end a small frame BB'. In BB' a horizontal axis works, at right angles to which is a small cylinder D, with a tightening screw H, through which passes a long rod GG', to one end of which is affixed a large ring AA', and along which slides a small cylinder carrying a weight W, which is capable of being fixed at any point of the

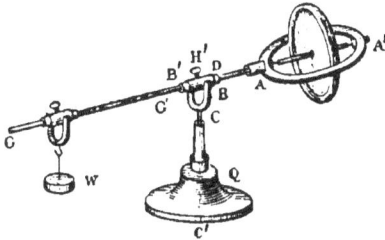

Fig. 71.

rod ; and so that it may act as a counterpoise to the ring, or to the ring and any weight attached to it.   An axis AA' works on pivots in the ring, in the same straight line with GG' ; to AA' a disc, or sphere, or cone, or any other body, can be attached, and thus can rotate about AA' as its axis ; to the body thus attached to AA' a rapid rotation can be given, either by means of a string wound round AA' or by a machine contrived for the purpose when AA' and its attached body are applied to it.   It is evident that the counterpoise W can be so adjusted that the centre of gravity of the rod, the ring, the attached body, and the counterpoise, should be in the axis BB' ; or at any point on either side of it ; that is, h may be positive, or be equal to o, or may be negative.   Also by fixing BB' in the arm of CC' which carries it, the inclination of the rod GG' to the vertical may be made constant ; that is, θ may be equal to θ_0 throughout the motion.   When the counterpoise is so adjusted that the centre of gravity of the rod GG' and its appendages is in CC', then $h = \text{o}$, or, what is equivalent, $mhg = \text{o}$."   (Price, *Calculus ;* vol. iv.)

It is evident that with such an instrument, with its various adjustments, all the motions about a fixed point can be fully displayed and examined; and the results already obtained in the case of the top (Art. 66) and the gyroscope (Art. 72) thereby shown.

91. Another form of gyroscope worthy of notice is that first constructed by Professor Gustav Magnus of Berlin, and described by him in Poggendorff's *Annalen der Physik und Chemie*, vol. xci., pp. 295–299. The instrument consists of two rings and discs such as AA', Fig. 71, connected by a rod supported in much the same way as the rod GG' in Fessel's gyroscope. There is a binding-screw at B, to arrest, when so desired, motion about the horizontal axis BB', and also a short rod projecting horizontally from the upper part of the vertical axis CC' by which the motion about that axis may be accelerated, retarded, or completely arrested at will. By means of two cords wound round their axes and simultaneously pulled off, the discs can be put in rapid rotation with nearly equal velocities either in the same or in opposite directions. The following phenomena are exhibited by this apparatus:

If the connecting rod be supported midway between the discs, and if the discs be made to rotate rapidly with equal velocities in the same direction, and no weight be suspended at W (Fig. 71), the connecting rod will remain at rest. If a weight be suspended at W, the rod and discs will slowly rotate about the vertical axis CC'. If the motion round the vertical axis be accelerated, the loaded end of GG' will rise, if the horizontal rotation be retarded, the loaded end will sink. If the binding-screw be tightened so as to arrest this rising or sinking, the rotation about the vertical axis will also cease, to commence again as soon as the binding-screw is loosened.

If the discs rotate with equal velocities in opposite directions, the loaded end of GG' will sink. If the connecting-rod be supported at a point nearer to one disc than to the other, and the discs be made to rotate with equal velocities in opposite directions, the instrument will still be found extremely sensitive.

NOTE ON THE PENDULUM AND THE TOP.

1. In Art. 35, pp. 47 to 49, we have found the equation

$$(h^2 + k^2)\left(\frac{d\theta}{dt}\right)^2 = 2gh(\cos\theta - \cos\alpha),$$

or, as it may be written (see page 50),

$$l\left(\frac{d\theta}{dt}\right)^2 = 2g(\cos\theta - \cos\alpha) \qquad (i)$$

for the oscillations of a rigid body about a fixed horizontal axis, and have applied it to the case of a pendulum making extremely small oscillations. We shall here consider the general case, when the arc of the oscillations is not necessarily small.

Let
$$\cos\theta - \cos\alpha = (1 - \cos\alpha)\cos^2\phi.$$

$$\therefore\; 1 - \cos\theta = (1 - \cos\alpha)\sin^2\phi$$

and
$$\cos\theta = \cos^2\phi + \cos\alpha\sin^2\phi. \qquad (ii)$$

Differentiating,

$$\sin\theta\frac{d\theta}{dt} = 2\sin\phi\cos\phi(1 - \cos\alpha)\frac{d\phi}{dt}.$$

$$\therefore\; (1 + \cos\theta)\left(\frac{d\theta}{dt}\right)^2 = 4(\cos\theta - \cos\alpha)\left(\frac{d\phi}{dt}\right)^2.$$

Substituting in (i),

$$l\left(\frac{d\phi}{dt}\right)^2 = g(1 - \sin^2\tfrac{1}{2}\alpha\sin^2\phi).$$

172

Let $\qquad\kappa^2 = \sin^2 \tfrac{1}{2}\,\alpha,$

and $\qquad l\nu^2 = g.$

$$\therefore \left(\frac{d\phi}{dt}\right)^2 = \nu^2(1 - \kappa^2 \sin^2 \phi).$$

$$\therefore \nu t = \int_0^\phi \frac{d\phi}{\sqrt{(1 - \kappa^2 \sin^2 \phi)}}, \qquad \text{(iii)}$$

an elliptic integral of the first kind.

$$\therefore \phi = \operatorname{am}(\nu t),$$

and (ii) becomes

$$\cos \theta = \operatorname{cn}^2(\nu t) + \cos \alpha \, \operatorname{sn}^2(\nu t). \qquad \text{(iv)}$$

Equation (ii) may be written in the form

$$\sin \tfrac{1}{2}\,\theta = \sin \tfrac{1}{2}\,\alpha \sin \phi\,;$$

consequently (iv) may be written in the form

$$\sin \tfrac{1}{2}\,\theta = \sin \tfrac{1}{2}\,\alpha \operatorname{sn} \nu t.$$

This equation determines the position of the pendulum at any given instant, and, by inversion, the times at which the pendulum is in a given position.

If T be the period of the pendulum, $i.e.$ the length of time required for the pendulum to make a double swing through the arc 2α,

$$\nu T = 4 \int_0^{\frac{\pi}{2}} \frac{d\phi}{\sqrt{(1 - \kappa^2 \sin^2 \phi)}}.$$

Integrating and writing $\frac{g}{l}$ for ν^2 and $\sin \tfrac{1}{2}\,\alpha$ for κ,

$$T = 2\pi \sqrt{\left(\frac{l}{g}\right)} \left\{ 1 + (\tfrac{1}{2})^2 (\sin \tfrac{1}{2})^2 + \left(\frac{1 \cdot 3}{2 \cdot 4}\right)^2 (\sin \tfrac{1}{2}\,\alpha)^4 \right.$$

$$\left. + \left(\frac{1 \cdot 3 \cdot 5}{2 \cdot 4 \cdot 6}\right)^2 (\sin \tfrac{1}{2}\,\alpha)^6 + \cdots \right\}. \qquad \text{(v)}$$

2. In Art. 67, p. 118, we have found the equation

$$\left(A \sin \theta \frac{d\theta}{dt}\right)^2 = (\cos \theta_0 - \cos \theta)\{2\, AMgh \sin^2 \theta$$
$$- C^2 n^2 (\cos \theta_0 - \cos \theta)\}, \qquad (a)$$

for the nutation oscillations of a top spinning about a fixed point, and in Art. 68 we have determined the approximate period of small oscillations. The period of oscillations of any magnitude and the value of θ at any given instant may be determined as follows:

Let $\qquad\qquad A = M (h^2 + k^2) = Mhl,$ (see page 50)

and $\qquad 2\, AMgh \sin^2 \theta - C^2 n^2 (\cos \theta_0 - \cos \theta)$

$$= 2\, AMgh (\cos \theta - \cos \theta_1)(\cosh \gamma - \cos \theta),$$

which requires that
$$\cos \theta_1 + \cosh \gamma = \frac{C^2 n^2}{2\, AMgh},$$

and $\qquad\qquad \cos \theta_1 \cdot \cosh \gamma = \frac{C^2 n^2 \cos \theta_0}{2\, AMgh} - 1.$

Substituting in (a), that equation becomes

$$l\left(\sin \theta \frac{d\theta}{dt}\right)^2 = 2\, g (\cos \theta_0 - \cos \theta)(\cos \theta - \cos \theta_1)(\cosh \gamma - \cos \theta). \quad (1)$$

Let $\quad \cos \theta_0 - \cos \theta = (\cos \theta_0 - \cos \theta_1) \cos^2 \tau.$

$\therefore\ \cos \theta - \cos \theta_1 = (\cos \theta_0 - \cos \theta_1) \sin^2 \tau,$

and $\qquad\qquad \cos \theta = \cos \theta_1 \cos^2 \tau + \cos \theta_0 \sin^2 \tau.$ $\qquad\qquad$ (2)

Differentiating,
$$- \sin \theta \frac{d\theta}{dt} = 2 \sin \tau \cos \tau (\cos \theta_0 - \cos \theta_1) \frac{d\tau}{dt}.$$

$$\therefore\ \left(\sin \theta \frac{d\theta}{dt}\right)^2 = 4 (\cos \theta_0 - \cos \theta)(\cos \theta - \cos \theta_1)\left(\frac{d\tau}{dt}\right)^2.$$

Substituting in (1),

$$l\left(\frac{d\tau}{dt}\right) = \tfrac{1}{2} g \{\cosh \gamma - \cos \theta_1 - (\cos \theta_0 - \cos \theta_1) \sin^2 \tau\}$$

$$= \tfrac{1}{2} g (\cosh \gamma - \cos \theta_1) \left\{ 1 - \frac{\cos \theta_0 - \cos \theta_1}{\cosh \gamma - \cos \theta_1} \sin^2 \tau \right\}$$

$$= g (\cosh^2 \tfrac{1}{2} \gamma - \cos^2 \tfrac{1}{2} \theta_1) \left\{ 1 - \frac{\cos^2 \tfrac{1}{2} \theta_0 - \cos^2 \tfrac{1}{2} \theta_1}{\cosh^2 \tfrac{1}{2} \gamma - \cos^2 \tfrac{1}{2} \theta_1} \sin^2 \tau \right\}.$$

Let
$$\kappa^2 = \frac{\cos^2 \tfrac{1}{2} \theta_0 - \cos^2 \tfrac{1}{2} \theta_1}{\cosh^2 \tfrac{1}{2} \gamma - \cos^2 \tfrac{1}{2} \theta_1},$$

and
$$l\nu^2 = g (\cosh^2 \tfrac{1}{2} \gamma - \cos^2 \tfrac{1}{2} \theta_1).$$

$$\therefore \left(\frac{d\tau}{dt} \right)^2 = \nu^2 (1 - \kappa^2 \sin^2 \tau).$$

$$\therefore \nu t = \int_0^\tau \frac{d\tau}{\sqrt{(1 - \kappa^2 \sin^2 \tau)}},$$

an elliptic integral of the first kind.

$$\therefore \tau = \operatorname{am}(\nu t),$$

and (2) becomes

$$\cos \theta = \cos \theta_1 \operatorname{cn}^2(\nu t) + \cos \theta_0 \operatorname{sn}^2(\nu t), \qquad (4)$$

thus determining the inclination of the axis of the top to the vertical at any given instant.

The period of a complete oscillation will be

$$T = \frac{4}{\nu} \int_0^{\frac{\pi}{2}} \frac{d\tau}{\sqrt{(1 - \kappa^2 \sin^2 \tau)}}$$

$$= 2\pi \sqrt{\left(\frac{l}{g (\cosh^2 \tfrac{1}{2} \gamma - \cos^2 \tfrac{1}{2} \theta_1)} \right)} \left\{ 1 + (\tfrac{1}{2})^2 \kappa^2 + \left(\frac{1 \cdot 3}{2 \cdot 4} \right)^2 \kappa^4 \right.$$

$$\left. + \left(\frac{1 \cdot 3 \cdot 5}{2 \cdot 4 \cdot 6} \right)^2 \kappa^6 + \cdots \right\}. \qquad (5)$$

Comparing equations (4) and (5) with (iv) and (v), it will be seen that the top's oscillations in nutation are of exactly the same character as the oscillations of an ordinary pendulum. Note, however, that in the discussions of the oscillations of the pendulum, θ is measured from an initial axis directed straight downwards, while in the discussion of the motion of the top, θ

is measured from the vertical axis as initial, so that θ, α, and ϕ in the former discussion should be replaced by $\pi-\theta$, $\pi-\alpha$, and $\pi-\tau$, to bring it into strict conformity of notation with the discussion of the movements of the top. On making these changes, it will be found that the pendulum oscillating about a fixed horizontal axis is merely the special case of the top in which $\theta_0=\pi-\alpha$, $\theta_1=\pi$, $n=0$, and, therefore, $\gamma=0$ and ψ is constant.

Equation (4) enables us to find the value of θ at any given instant t, but to completely determine the position of the top, it is necessary to be able also to find ψ. To do this requires the integration of the equation (b) of Art. 65, p. 116, which may be reduced to the integration of two elliptic integrals of the third kind, as follows:

$$A \sin^2\theta \frac{d\psi}{dt} = Cn(\cos\theta_0 - \cos\theta). \tag{b}$$

$$\therefore \frac{d\psi}{dt} = \frac{Cn}{A}\left(\frac{\cos^2\frac{1}{2}\theta_0}{1+\cos\theta} - \frac{\sin^2\frac{1}{2}\theta_0}{1-\cos\theta}\right)$$

$$= \frac{Cn}{A}\left(\frac{\cos^2\frac{1}{2}\theta_0}{1+\cos\theta_1\cos^2\tau+\cos\theta_0\sin^2\tau}\right.$$

$$\left.- \frac{\sin^2\frac{1}{2}\theta_0}{1-\cos\theta_1\cos^2\tau-\cos\theta_0\sin^2\tau}\right)$$

$$= \frac{Cn}{A}\left(\frac{\cos^2\frac{1}{2}\theta_0}{1+\cos\theta_1+(\cos\theta_0-\cos\theta_1)\operatorname{sn}^2(\nu t)}\right.$$

$$\left.- \frac{\sin^2\frac{1}{2}\theta_0}{1-\cos\theta_1-(\cos\theta_0-\cos\theta_1)\operatorname{sn}^2(\nu t)}\right).$$

Thus ψ is expressed as the difference of two elliptic integrals of the third kind.

MISCELLANEOUS EXAMPLES.

———∞o⋅ಿⲟ∞———

1. Find the principal axes of a quadrant of an ellipse at the centre.

2. If a rigid body be referred to three rectangular axes such that $A = B$ and $\Sigma(mxy) = 0$, show that the mean principal moment of inertia $= A$.

3. Determine the position of a point O in a triangular lamina, such that the moments of inertia of AOB, BOC, COA, about an axis through O, perpendicular to the plane of the lamina, may all be equal.

4. Find the moment of inertia of the solid formed by the revolution of the curve $r = a(1 + \cos\theta)$ about the initial line, about a line through the pole perpendicular to the initial line.

5. A uniform wire is bent into the form of a catenary. Find its moments of inertia about its axis, and its directrix.

6. Find the moment of inertia of a paraboloid of revolution about a tangent line at the vertex; the density in any plane perpendicular to the axis varying as the inverse fifth power of its distance from the vertex.

7. Find the moment of inertia of a semi-ellipse cut off by the axis minor about the line joining the focus with the extremity of the axis minor.

8. If the moments of inertia of a rigid body about three axes, passing through a point and mutually at right angles, be equal to one another, show that these axes are on the surface of an

elliptic cone whose axis is that of least or greatest moment according as the mean moment of inertia is greater or less than the arithmetic mean between the other two.

9. Show that the moment of inertia of a regular octahedron about one of its edges is $\frac{3}{5} a^2 M$, where a is the length of an edge and M is the mass of the octahedron.

10. Prove that the moment of inertia of a solid regular tetrahedron about any axis through its centre of inertia is $M\frac{a^2}{20}$, a being the length of an edge.

11. If β, γ be the perpendiculars from B and C on a principal axis at the angular point A of the triangle ABC, show that

$$(a^2 - b^2 - c^2)(\beta^2 - \gamma^2) = b^2\beta^2 + c^2\gamma^2 + 2(b^2 - c^2)\beta\gamma.$$

12. Show that if a plane figure have the moments of inertia round two lines in it, not perpendicular to one another, equal, a principal axis with respect to the point of intersection bisects the angle between them.

13. Determine the points of an oblate spheroid with respect to which the three principal moments are equal to one another.

14. Show that the conditions which must be satisfied by a given straight line in order that it may, at some point of its length, be a principal axis of a given rigid body, is always satisfied if the rigid body be a lamina and the straight line be in its plane, unless the straight line pass through the centre of inertia.

15. If a straight line be a principal axis of a rigid body at every point in its length, it must pass through the centre of inertia of body.

16. Assuming that the radius of gyration of a regular polygon of n sides about any axis through its centre of inertia and in its own plane, is

$$\frac{c}{2\sqrt{b}}\sqrt{\left\{\left(2+\cos\frac{2\pi}{n}\right)\bigg/\left(1-\cos\frac{2\pi}{n}\right)\right\}},$$

where c is the length of any side; find the radius of gyration of a circular disc about a line through any point in its circumference and perpendicular to its plane, and show that it is equal to the radius of a circular ring about a tangent.

17. The locus of a point such that the sum of the moments of inertia about the principal axes through the point is constant, is a sphere whose centre is the centre of inertia of the body.

18. If A, B, C be the moments of inertia of a body about principal axes,

$$A \cos^2 \alpha + B \cos^2 \beta + C \cos^2 \gamma$$

will be the moment of inertia about any other axis passing through the origin and having $\cos a \cos \beta \cos \gamma$ for its direction-cosines.

19. If the centre of inertia of a rigid body be the origin, and the principal axes at that point the axes of coördinates, then at an umbilicus of the ellipsoid

$$\frac{x^2}{A+\lambda} + \frac{y^2}{B+\lambda} + \frac{z^2}{C+\lambda} = 1,$$

two of the principal moments of inertia will be equal.

20. If the density at any point of a right circular cone be proportional to the distance from the exterior surface, show that the radius of gyration about the axis of figure is $\frac{a^2}{10}$, where a is the radius of the base.

21. Find the moment of inertia of the solid

$$(x^2 + y^2 + z^2 - ax)^2 = a^2(x^2 + y^2 + z^2)$$

about the axis of x.

Find also the moment of inertia of the surface of this solid.

22. The locus of points at which one of the principal axes passes through a fixed point in one of the principal planes through the centre of inertia, is a circle.

23. If a and b be the sides of a homogeneous parallelogram, θ and ϕ the inclinations of its principal axes in its own plane,

through its centre of inertia, to these sides respectively, show
that $a^2 \sin 2\,\theta = b^2 \sin 2\,\phi.$

24. Find the moments of inertia of a uniform circular lamina about its principal axes through a given point in its plane.

25. Show that two of the principal moments of inertia with respect to a point in a rigid body cannot be equal unless two are equal with respect to the centre of inertia and the point be situated on the axis of unequal moment.

26. Prove that in any rigid body the locus of the point through which one of the principal axes is in a given direction is a rectangular hyperbola whose plane passes through the centre of inertia, and one of whose asymptotes is in the given direction; unless the given direction be that of one of the principal axes through the centre of inertia.

27. A series of parabolas are described in one plane having a common vertex A and a common axis, and from a point P in one of them an ordinate PN is drawn to the axis. Show that if the moment of inertia of the curvilinear area APN about an axis through A perpendicular to the plane of the parabolas be proportional to the area APN, the locus of P is an ellipse.

28. Show that if the momental ellipsoid at a point not in one of the principal planes through the centre of inertia be a spheroid, it will at the centre of inertia be a sphere.

29. Find the moment of inertia of a segment of a circle about its chord.

30. Find the moment of inertia of an equilateral triangular lamina about an axis through the centre of inertia and perpendicular to the lamina if the density of the lamina at any point varies directly as the distance of the point from the centre of inertia.

31. If A, B, C be the moments of inertia about principal axes through the centre of inertia and α, β, γ be the moments of inertia about principal axes through a point P, show that

(I) If $(\alpha + \beta - \gamma) = A + B - C$, the locus of P will be one of the principal planes through the centre of inertia.

(II) If $\alpha + \beta + \gamma$ be constant, the locus of P will be a sphere with centre at centre of inertia.

(III) If
$$(\sqrt{\alpha} + \sqrt{\beta} + \sqrt{\gamma})(\sqrt{\beta} + \sqrt{\gamma} - \sqrt{\alpha})$$
$$\times (\sqrt{\gamma} + \sqrt{\alpha} - \sqrt{\beta})(\sqrt{\alpha} + \sqrt{\beta} - \sqrt{\gamma})$$

be constant, the locus of P will be an ellipsoid similar and similarly situated and concentric with the central ellipsoid at the centre of inertia.

(IV) If $\beta - \gamma < \alpha$ and P lie on a lemniscate of revolution having for foci the points where the momental ellipsoid is a sphere, $\alpha - \beta = A - B$, α and β being the moments about the axes through P which pass through the axis A.

32. Find the moment of inertia of a portion of the arc of an equiangular spiral about a line through its pole perpendicular to its plane.

33. Find the moment of inertia of the segment of a parabolic area bounded by a chord perpendicular to its axis, about any line in its plane through the focus ; and determine the position of the chord that all such moments may be equal.

34. Prove that if the height of a homogeneous right circular cylinder be to its diameter as $\sqrt{3} : 2$, the moments of inertia of the cylinder about all axes passing through the centre of inertia will be equal.

35. Find the moment of inertia of a parabolic area bounded by the latus rectum about the line joining its vertex to the extremity of its latus rectum.

36. Find the locus of those diameters of an ellipsoid, the moments of inertia about which are equal to the moment of inertia about the mean axis.

37. One extremity of a string is attached to a fixed point; the string passes round a rough pulley of given radius and over a

smooth peg and is attached to a weight equal to that of the pulley. Determine the motion, the positions of the string on either side of the pulley being vertical.

38. A heavy uniform rod has at one extremity a ring which slides on a smooth vertical axis; the other extremity is in motion on a horizontal plane, and is connected by an elastic string with the point where the axis meets the plane. Determine the motion supposing the string always stretched.

39. A ball spinning about a vertical axis moves on a smooth horizontal table and impinges directly on a perfectly rough vertical cushion. Show that the *vis viva* of the ball is diminished in the ratio $10 e^2 + 14 \tan^2 \theta : 10 + 49 \tan^2 \theta$, where e is the co-efficient of restitution of the ball and θ the angle of reflection.

40. A homogeneous lamina rotating in its plane about its centre of inertia, is brought suddenly to rest by sticking a two-pronged fork into it. Show that the impulses on the prongs are equal to one another, and are of the same magnitude wherever the fork is stuck in.

41. A free rod is at rest and a ball is fired at it to break it. Show that it will be most likely to cause it to break if it strike it at the midpoint, or at one-sixth of its length from either end; and that it will be least likely to break the rod if it strike it at one-third of its length from either end. And that in either case the most likely point for it to snap is the middle point.

42. Three pieces cut from the same uniform rigid wire are connected together so as to form a triangle ABC, which is then set in contact with a smooth horizontal plane. Find the direction and magnitude of the strains at the angular connections.

Prove the following construction for the direction of the strains: If AB, AC be produced to D, E respectively, and BD and CE be each made equal to BC, then will DE be parallel to the direction of the strain at A.

Also show that the direction of the strain at A makes with the side BC an angle

$$= \tan^{-1}\left\{\frac{\sin\beta \sim \sin C}{1 + \cos\beta + \cos C}\right\}.$$

43. The ends of a uniform heavy rod move on the same smooth, fixed, vertical ring. Determine its angular velocity in the lowest position, supposing it to fall from a given position starting from rest.

44. A body revolves about a horizontal axis, starting from rest when the centre of inertia is in the horizontal plane containing the axis. Show that when the body has revolved through 45°, the effective force upon the centre of inertia makes with the vertical an angle $= \tan^{-1} 3$.

45. A box is fixed upon a horizontal plane and its lid is placed in a vertical position; a blow is given to the lid at the midpoint of the upper edge and perpendicular to its plane. Determine the initial impulse on the hinges, the finite pressure on them during the motion, and the impulsive pressure on them when the lid impinges on the opposite edge and closes the box.

46. AB, BC are two equal heavy rods hinged together at B; the rod AB is capable of moving in a vertical plane about A, and C can slide by means of a small ring along a vertical axis passing through A. Find the angular velocity with which the whole must revolve about AC that the triangle ABC may be equilateral.

47. A string with one end fastened to a smooth vertical wall is wrapped round a cylinder which is then placed in contact with the wall. Find the velocity of the cylinder and the tension of the string in terms of the inclination of the string to the wall.

48. The lower extremity of a heavy uniform beam of length a slides on a weightless inextensible string of length $2a$, whose extremities are attached to two fixed points in a horizontal line, and the upper extremity slides on a vertical rod which bisects the line joining the fixed points. Prove that the only position

of equilibrium of the beam is vertical and that the time of a small oscillation about this position is $\dfrac{2\pi a}{\sqrt{\{3g(a\sim 2b)\}}}$, where $2\sqrt{(a^2-b^2)}$ is the distance between the two fixed points.

49. P pulls Q by means of an unextensible string passing over a rough pulley in the form of a vertical circle, which can turn freely about an axis through its centre, which is fixed. Determine the velocity attained after a given space has been described.

50. A hoop of mass M rolls down a rough inclined plane, and carries a heavy particle of mass m at a point of its circumference. Determine the motion.

51. A hollow tubular ring of radius a contains a heavy particle with its plain vertical upon a smooth horizontal plane; a horizontal velocity $2\sqrt{2ag}$ is communicated to the ring in its own plane. Show that the particle will just rise to the top of the tube.

52. Four equal particles are connected by four equal strings, which form a square, and the particles repel each other with a force varying directly as the distance. If one of the strings be cut, find the velocity of each particle at the instant when they are all in a straight line.

53. One half of the inner surface of a fixed hemispherical bowl is smooth and the other half rough; a solid sphere slides down the smooth part of the bowl, starting from rest at the horizontal rim, and at the bottom comes in contact with and rolls up the rough part of the surface. Find the change of *vis viva* of the sphere at the bottom of the bowl, and show that if θ be the angle which the line joining the centres of the sphere and bowl makes with the vertical when the sphere begins to descend the rough surface, $\cos\theta=\frac{2}{7}$.

54. A cone of mass m and vertical angle 2α can move freely about its axis and has a fine smooth groove cut along its surface so as to make a constant angle B with the generating

lines of the cone. A heavy particle of mass P moves along the groove under the action of gravity, the system being initially at rest with the particle at a distance c from the vertex. Show that if θ be the angle through which the cone has turned when the particle is at any distance r from the vertex, then

$$\frac{mk^2 + Pr^2 \sin^2 \alpha}{mk^2 + Pc^2 \sin^2 \alpha} = 2\,\theta \sin \alpha \cot \beta e,$$

k being the radius of gyration of the cone about its axis.

55. A heavy ring just fitting round a smooth vertical cylinder is suspended by n vertical strings of equal lengths, and fixed to the ring at equidistant points. When an angular velocity is given to the ring about its centre, show that the height to which it rises is independent of the length of the strings. Find also the greatest value of the angle through which it turns.

56. A sphere has a fine wire fastened normally to a point on its surface, the other end being fastened to a point on a rough inclined plane. If the sphere be slightly displaced from its position of equilibrium on the plane, find the time of a small oscillation, neglecting the weight of the wire.

57. Two equal uniform beams AB, AC are freely movable in a vertical plane about A, B and C are connected by an elastic string whose natural length is equal to AB. The beams are held in a vertical position and suffered to descend. Determine the motion, the coefficient of elasticity of the string being equal to four times the weight of either beam.

58. A circular wire is revolving uniformly about its centre fixed. If it be cracked at any point, show that the tendency to break at an angular distance α from the crack is proportional to $\sin^2 \frac{\alpha}{2}$.

59. A disc which has a particle of equal mass attached to its circumference, rolls on a rough inclined plane. Determine the motion and the friction in any position of the disc, supposing it

to start from the position in which the particle is in contact
with the plane.

60. A spherical shell whose centre is fixed contains a rough
ball which is held at one extremity of a horizontal diameter of
the shell and then allowed to descend. If the radius of the
shell be three times that of the ball, and when the ball is next in
instantaneous rest the same point of each is again in contact, the
angular velocity of the line joining their centres is $3\sqrt{\left\{\frac{g\sin\theta}{10\,a}\right\}}$,
θ being the inclination of this line to the horizon, and a being
the radius of the ball.

61. A circular ring is suspended with its plane horizontal, by
three equal vertical inextensible strings attached at equal distances to its circumference. If the ring be twisted till the
strings just meet in a point, and be then left to itself, find its
angular velocity when the strings are vertical again.

62. Two rods AB, BC connected·by a hinge at B are in
motion on a smooth horizontal plane, the end A being fixed.
If initially AB has no angular velocity, that of BC being ω, show
that when BC has no angular velocity, that of AB will be $\dfrac{b\omega}{2\,a}$
and the angle between the rods will be $\cos^{-1}\left\{\dfrac{4(b-2\,a)}{3\,b}\right\}$, $2\,a$
and $2\,b$ being the lengths of the rods which are supposed equal
in mass.

63. A uniform heavy beam of length $2\,c$ is supported in a
horizontal position by means of two strings without weight, each
of length b, which are fastened to its ends, the other ends of the
strings being fixed; in equilibrium each of the strings makes an
angle u with the horizontal. Find the time of a small oscillation
when the system is slightly displaced in the vertical plane in
which it is situated, the strings not being slackened.

64. A lamina bounded by a cycloid and its base has its centre
of inertia at the middle point of its axis. It is placed with its
base vertical on a perfectly rough horizontal plane, and allowed

to roll down. Show that at the moment its vertex reaches the plane its angular velocity is $\sqrt{\left\{\dfrac{2\,ag(\pi-1)}{a^2+k^2}\right\}}$, where a is the radius of the generating circle and k the radius of gyration about the centre of inertia.

65. A wire is bent into the form of the lemniscate $r^2=a^2\cos 2\theta$, and laid upon a smooth horizontal table; a fly walks along the top of the wire, starting from one vertex. Show that if the masses of the wire and fly be in the ratio $a^2:k^2$, where k is the radius of gyration of the lemniscate about a vertical axis through the node, then when the fly has arrived at the node the wire has turned through an angle $\dfrac{\pi}{4}-\dfrac{1}{2}$.

66. A uniform circular wire of radius a, movable about a fixed point in its circumference, lies on a smooth horizontal plane. An insect of mass, equal to that of the wire, crawls along it, starting from the extremity of the diameter opposite to the fixed point, its velocity relative to the wire being uniform and equal to v. Prove that after a time t the wire will have turned through an angle $\dfrac{vt}{2a}-\dfrac{1}{\sqrt 3}\tan^{-1}\left(\dfrac{1}{\sqrt 3}\tan\dfrac{vt}{2a}\right)$.

67. A uniform string is stretched along a smooth inclined plane which rests on a smooth horizontal table. Enough of the string hangs over the top of the plane to keep the whole system at rest. If the string be gently pulled over the plane, and the whole system be then left to itself, investigate the ensuing motion, supposing the length of the string to be equal to the height of the plane.

68. Two particles of equal mass are attached to the extremities of a rigid rod without inertia, movable in all directions about its middle point. The rod being set in motion from a given position with given velocity, find equations to determine its subsequent motion.

69. A rod of length $2a$ movable about its lower end is inclined at an angle α to the vertical, and is given a rotation ω about the

vertical. If θ be its inclination to the vertical when its angular velocity about a horizontal axis is a maximum, show that

$$3\,g\sin^3\theta\tan\theta + 4\,a\omega^2\sin^4\alpha = 0.$$

70. The time of descent, down a rough inclined plane, of a spherical shell which contains a smooth solid sphere of the same material as itself is t_1, the time of descent down the same plane of a solid sphere of the same material and radius as the shell is t_2. Determine the thickness of the shell.

71. A heavy chain, flexible, inextensible, homogeneous, and smooth, hangs over a small pulley at the common vertex of two smooth inclined planes. Apply d'Alembert's principle to determine the motion of the chain.

72. A perfectly rough right prism, whose section is a square, is placed with its axis horizontal upon a board of equal mass lying on a smooth horizontal table. A vertical plane containing the centres of inertia of the two is perpendicular to the axis of the prism; a horizontal blow in this plane communicates motion to the system. Show that the prism will topple over if the momentum of the blow be greater than that acquired by the system falling through a height $\dfrac{13}{12}\tan\dfrac{\pi}{8}a$, where a is a side of the square section of the prism.

73. Determine the small oscillations in space of a uniform heavy rod of length $2a$, suspended from a fixed point by an inextensible string of length l fastened to one extremity. Prove that if x be one of the horizontal coördinates of that extremity of the rod to which the string is fastened

$$x = A \sin (n_1 t + \alpha) + B \sin (n_2 t + \beta),$$

where n_1, n_2 are the two positive roots of the equation

$$aln^4 - (4a + 3l)gn^2 + 3g^2 = 0$$

and A, B, α, β are arbitrary constants.

74. The bore of a gun-barrel is formed by the motion of an ellipse whose centre is in the axis of the barrel and plane perpendicular to that axis, the centre moving along the axis and the ellipse revolving in its own plane with an angular velocity always bearing the same ratio to the linear velocity of its centre. A spheroidal ball fitting the barrel is fired from the gun. If v be the velocity with which the ball would have emerged from the barrel had there been no twist, prove that the velocity of rotation with which it actually emerges in the case supposed is

$$\frac{2 \pi n v}{\sqrt{(l^2 + 4 \pi^2 n^2 k^2)}},$$

the number of revolutions of the ellipse corresponding to the whole length l of the barrel being n, and k being the radius of gyration of the ball about the axis coinciding with the axis of the barrel, and the gun being supposed to be immovable.

75. A plane lamina moving either about a fixed axis or instantaneously about a principal axis, impinges on a free inelastic particle in the line through the centre of inertia of the lamina perpendicular to the axis of rotation at the moment of impact. If the velocity of the particle after impact be the maximum velocity, prove that the angular velocity of the lamina will be diminished in the ratio of $1 : 2$.

76. Two equal uniform rods are placed in the form of the letter X on a smooth horizontal plane, the upper and the lower extremities being connected by equal strings. Show that whichever string be cut the tension of the other will be the same function of the rods, and initially is $\frac{3}{8} g \sin \alpha$, where α is the inclination of the rods.

77. An equilateral triangle is suspended from a point by three strings, each equal to one of the sides, attached to its angular points. If one of the strings be cut, show that the tensions of the other two are diminished in the ratio of $36 : 43$.

78. Apply *the principle of energy* to determine the time of a small oscillation of a uniform rod placed in a smooth, fixed, hemispherical bowl, the motion taking place in a vertical plane.

79. A frame formed of four equal uniform rods loosely jointed together at the angular points, so as to form a rhombus, is laid on a smooth horizontal plane and a blow is applied to one of the rods in a direction at right angles to it. Prove that the frame will begin to move as a rigid body provided the middle point of the rod which receives the blow be equidistant from the line of action of the blow and the perpendicular dropped upon the rod from the centre of inertia of the frame.

Prove also that in this case the initial angular velocity of the rod which receives the blow is one-eighth of what it would have been had it been unconnected with the remaining rods.

80. Three equal uniform rods AB, BC, CD, freely jointed at B and C, are lying in one straight line on a smooth horizontal table, and an impulse is applied at the midpoint of BC, perpendicular to that rod. Find the stresses on the hinges at B and C in any subsequent positions of the rods, and show that when AB, CD are perpendicular to BC, their midpoints are moving in directions which make an angle $\cos^{-1}(\frac{1}{3})$ with BC.

81. A parallelogram is formed of four rigid uniform rods freely jointed at their extremities. If the parallelogram be laid on a smooth horizontal table and a blow be applied to any one of the rods at right angles to it, and in a direction passing through the intersection of the lines drawn through its extremities parallel to the diagonals, determine the initial motion of the parallelogram.

82. A circular disc is capable of motion about a horizontal tangent which rotates with uniform angular velocity ω about a fixed vertical axis through the point of contact. Prove that if the disc be inclined at a constant angle a to the horizontal,

$$\omega^2 \sin a = \frac{4g}{5a}.$$

83. A uniform rod is rotating with angular velocity $\sqrt{\left(\dfrac{8g}{4a}\right)}$ about its centre of inertia, which is at rest at the instant when the rod, being vertical, comes in contact with an inelastic plane inclined to the horizontal at an angle $\sin^{-1}\sqrt{\tfrac{1}{2}}$. The motion being in a vertical plane normal to the inclined plane, prove that the angular velocity of the rod when it leaves the inclined plane is $\sqrt{\left\{\dfrac{g}{a}\left(1+\dfrac{1}{\sqrt{2}}\right)\right\}}$.

84. If a rigid body which is initially at rest, and which has a point in it fixed, is struck by a given impulsive couple, show that the *vis viva* generated is greater than that which would have been generated by the same couple if the body had been constrained to turn about an axis through the fixed point and not coincident with the axis of spontaneous rotation.

85. A and B are two fixed points in the same horizontal line; CD, a heavy uniform rod equal in length to AB, is suspended by four inextensible strings AC, AD, BC, BD, where AC is equal in length to BD, and AD to BC. If two of the strings AC, BD be cut, determine the tension of the other two immediately after cutting, and find the angular velocity of the rod when it reaches its lowest position.

86. A beam AB is fixed at A. At B is fastened an elastic string whose natural length is equal to AB; the other end of the string is fastened to a point C vertically above A, AC being equal to AB. The beam is held vertically upwards and then displaced. If it come to rest when hanging vertically downwards, find the greatest pressure on the axis during the motion.

87. Two equal rods AB, BC are connected by a hinge at B. A is fixed and C is in contact with a smooth horizontal plane, the system being capable of motion in a vertical plane. If motion commence when the rods are inclined at an angle α to the horizon, show that there will be no pressure at the hinge when their inclination θ is given by the equation

$$3(\sin^3\theta + \sin\theta) = 2\sin\alpha.$$

88. A rod AB is movable freely in a vertical plane about A; to B is fastened an elastic string, the other end being attached to a point C in the vertical plane at such a distance from A that when the rod is held horizontal the tension on the string vanishes. If the rod be now allowed to fall, find the modulus of elasticity of the string that the rod may just reach a vertical position.

89. A prolate spheroid is fixed at one of its poles, and is allowed to fall from its position of unstable equilibrium under the action of gravity only. Find the pressure at the fixed point in any subsequent position.

90. Every particle of two equal uniform rods, each of length $2a$, attracts every other particle according to the law of gravitation; the rods are initially at right angles and are free to move in a plane about their midpoints, which are also their centres of inertia and are coincident. If angular velocities ω, ω' be communicated to the rods respectively, show that at the time t the angle θ between them is given by the equation

$$\left(\frac{d\theta}{dt}\right)^2 = (\omega - \omega')^2 + \frac{12\,m}{a^3}\log\frac{(3 - 2\sqrt{2})\left(\cos\frac{\theta}{2} + \sin\frac{\theta}{2} + 1\right)}{\cos\frac{\theta}{2} + \sin\frac{\theta}{2} - 1}.$$

91. Two equal spheres of radius a and mass M are attached to the extremities of a rigid rod of the same material, whose length is $4a$ and section $\frac{1}{70}$ of a principal section of the sphere. If the rod can move freely about its midpoint and one sphere be struck by a blow P normal to it and the rod, the time which must elapse before the other sphere takes the place of this one is

$$\frac{44\,\pi a M}{7\,P}.$$

92. A thin uniform rod, one end of which is attached to a smooth hinge, is allowed to fall from a horizontal position. Prove that the stress on the hinge in any given direction is a maximum when the rod is equally inclined to this direction and to the vertical, and the stress perpendicular to this is then

$\frac{11}{8} W \cos \alpha$, where W is the weight of the rod and α is the inclination of the given direction to the horizontal.

93. A man standing in a swing is set in motion. Supposing that the initial inclination of the swing to the vertical is given, and that the man always crouches when in the highest position, and stands up when in the lowest, find how much the arc of vibration will be increased after n complete oscillations.

94. Three rods are hinged together so as to form an isosceles triangle ABC, A being the vertex. The whole is rotating with angular velocity ω round an axis through the middle point of the base and perpendicular to the plane of the rods, when it is suddenly brought to rest. Show that the impulsive action at A bisects the angle BAC, and find its magnitude.

95. A triangular lamina is suspended at rest horizontally by vertical strings attached to its angular points A, B, C. If the strings at B and C be simultaneously cut, show that there will be no instantaneous change of tension in the string at A, if AD be perpendicular to either AB or AC. $AD = CD \cos ADC$, D being the midpoint of BC.

96. A hollow spherical shell is filled with homogeneous fluid which gradually solidifies without alteration of density, the solidification proceeding uniformly from the outer surface, so that the mass of the solidified portion is proportional to the time. If the shell initially rotate about a given axis with a given angular velocity ω, find the angular velocity at any subsequent period before the solidification is complete.

97. A lamina whose centre of gravity is G is revolving about a horizontal axis perpendicular to it and meeting it in C. Supposing it to begin to move from a position in which CG is horizontal, prove that the greatest angle which the direction of the pressure on the axis can make with the vertical is $\cot^{-1}\left(\dfrac{2}{3}\dfrac{k^2}{h^2}\tan\theta\right)$, where θ is the corresponding angle which CG

o

makes with the vertical, k is the radius of gyration about an axis through G perpendicular to the lamina, and $h = CG$.

98. A rough uniform rod, length $2a$, is placed with a length $c(>a)$ projecting over the edge of the table. Prove that the rod will begin to slide over the edge when it has turned through an angle $\tan^{-1} \dfrac{\mu a^2}{a^2 + 9(c-a)^2}$.

99. If gravity be the only force acting on a body capable of freely turning about a fixed axis and the body be started from its position of stable equilibrium with such a velocity that it may just reach its position of unstable equilibrium, find the time of describing any angle.

100. If an isosceles triangle move, under the action of gravity only, about its base as a fixed axis starting from a horizontal position, show that the greatest pressure on the axis is seven-thirds the weight.

101. If the centre of oscillation of a triangle, suspended from an angular point and oscillating with its plane vertical, lie on the side opposite the point of suspension, show that the angle at the point must be a right angle.

102. A horizontal circular tube of small section and given mass is freely movable about a vertical axis through its centre. A heavy particle within the tube is projected along it with a given velocity. Given the coefficient of friction between the tube and the particle, determine the terminal velocity of both, and the time which must elapse before that motion is attained.

103. Part of a heavy chain is coiled round a cylinder freely movable about its axis of figure which is horizontal, and the remainder hangs vertically. Determine the motion, supposing the system to start from rest and neglecting the thickness of the chain.

104. Two weights are connected by a fine chain which passes over a wheel free to rotate about its centre in a vertical plane.

Given the coefficient of friction between the string and the wheel, find the condition which determines whether the string will slide over the wheel or will not slide.

105. Two straight equal and uniform rods are connected at their ends by fine strings of equal length a so as to form a parallelogram. One rod is supported at its centre by a fixed axis about which it can turn freely, this axis being perpendicular to the plane of motion, which is vertical. Show that the middle point of the lower rod will oscillate in the same way as a simple pendulum of length a, and that the angular motion of the rods is independent of this oscillation.

106. A loaded cannon is suspended from a fixed horizontal axis, and rests with its axis horizontal and perpendicular to the fixed axis, the supporting ropes being equally inclined to the vertical. If v be the initial velocity of the ball whose mass is $\dfrac{1}{n}$ of the weight of the cannon, and h the distance between the axis of the cannon and the fixed axis of support, show that when the cannon is fired off the tension of each rope is immediately changed in the ratio $v^2 + n^2gh : n(n + 1)gh$.

107. Two equal triangles ABC, $A'B'C'$, right-angled at C and C', rotate about their equal sides CA and $A'C'$ as fixed axes in the same horizontal straight line. The distance CC' is less than the sum of the sides CA, $A'C'$. The triangles, being at first placed horizontally, impinge on one another when vertical. Determine the initial subsequent motion and discuss the case in which AA' is less than one-fifth CC'.

108. Find the envelope of all the axes of suspension that lie in a principal plane through the centre of inertia of a rigid body, and such that the length of the simple pendulum may be always twice the radius of gyration of the body about one of the axes lying in the plane.

109. A flat board bounded by two equal parabolas with their axis and foci coincident, and their concavities turned towards

each other, is capable of moving about the tangent at one of the vertices. Find the centre of percussion.

110. A uniform beam capable of motion about its middle point is in equilibrium in a horizontal position; a perfectly elastic ball, whose mass is one-fourth that of the beam, is dropped upon one extremity and is afterwards struck by the other extremity of the beam. Prove that the height from which the ball was dropped was $\frac{49}{48}(2n + 1)\pi \times$ length of beam.

111. Two equal circular discs are attached, each by a point in its circumference, to a horizontal axis, one of them in the plane of the axis and the other perpendicular to it, and each is struck by a horizontal blow which, without creating any shock on the axis, makes the disc revolve through 90°. Show that the two blows are as $\sqrt{6} : \sqrt{5}$.

112. A rigid body capable of rotation about a fixed axis is struck by a blow so that the axis sustains no impulse. Prove that the axis must be a principal axis of the body at the point where it is met by the perpendicular let fall on it from the point of application of the blow.

113. A uniform rod AB of mass M is freely movable about its extremity A, which is fixed; at C, a point such that AC is horizontal and equal to AB, a smooth peg is fixed over which passes an inelastic string fastened to the rod at B, and to a body also of mass M which is supported in a position also below C. If the rod be allowed to fall from coincidence with AC, and the string be of such a length as not to become tight until the rod is vertical, the angular velocity of the rod will be suddenly diminished by three-fifths.

114. A piece of wire is bent into the form of an isosceles triangle and revolves about an axis through its vertex perpendicular to its plane. Find the centre of oscillation and show that it will lie in the base when the triangle is equilateral.

115. A circular lamina performs small oscillations
(1) about a tangent line at a given point of its circumference,
(2) about a line through the same point perpendicular to its plane.

Compare the times of oscillation.

116. A uniform beam is drawn over the edge of a rough horizontal table so that only one-third of its length is in contact with the table; and it is then abandoned to the action of gravity. Show that it will begin to *slide* over the edge of the table when it has turned through an angle equal to $\tan^{-1}\frac{\mu}{2}$, μ being the coefficient of friction between the beam and the table.

117. A uniform beam AB, capable of motion about A, is in equilibrium. Find the point at which a blow must be applied in order that the impulse at A may be one-eighth of the blow.

118. A rectangle is struck by an impulse perpendicular to its plane. Determine the axis about which it will begin to revolve, and the position of this axis with reference to an ellipse inscribed in the rectangle.

119. A rectangle rotates about one side as a fixed axis. Find the pressure on the axis (1) when horizontal, (2) when inclined to the horizontal.

120. About what fixed axis will a given ellipsoid oscillate in the shortest possible time?

121. A uniform semicircular lamina rotates about a fixed horizontal axis through its centre in its plane. Determine the stresses on this axis.

122. If T_1 and T_2 are the times of a small oscillation of a rigid body, acted on only by gravity, about parallel axes which are distant a_1 and a_2 respectively from the centre of inertia, and T be the time of a small oscillation for a simple pendulum of length $a_1 + a_2$, then will $(a_1 - a_2)T^2 = a_1 T_1^2 - a_2 T_2^2$.

123. A uniform beam of mass m, capable of motion about its middle point, has attached to its extremities by strings, each of

length l, two particles, each of mass p, which hang freely. When the beam is in equilibrium, inclined at an angle α to the vertical, one of the strings is cut; prove that the initial tension of the other string is $\dfrac{mpg}{m + 3p\sin^2\alpha}$, and that the radius of curvature of the initial path of the particle is $\dfrac{9\,lp\sin^3\alpha}{m\cos\alpha}$.

124. A uniform inelastic beam capable of revolving about its centre of inertia, in a vertical plane, is inclined at an angle α to the horizontal, and a heavy particle is let fall upon it from a point in the horizontal plane through the upper extremity of the beam. Find the position of this point in order that the angular velocity generated may be a maximum.

125. A uniform elliptic board swings about a horizontal axis at right angles to the plane of the board and passing through one focus. Prove that if the excentricity of the ellipse be $\sqrt{\tfrac{2}{5}}$, the centre of oscillation will be the other focus.

126. A circular ring hangs in a vertical plane on two pegs. If one peg be removed, prove that, P_1, P_2 being the instantaneous pressures on the other peg calculated on the supposition that the ring is (1) smooth, (2) rough, $P_1{}^2 : P_2{}^2 :: 1 : 1 + \tfrac{1}{4}\tan^2\alpha$, where α is the angle which the line drawn from the centre of the ring to the centre of the peg makes with the vertical.

127. A uniform beam can rotate about a horizontal axis so placed that a ball of weight equal to that of the beam, resting on one end of the beam, keeps it horizontal. A blow, perpendicular to the length of the beam, is struck at the other end. Investigate the action between the ball and the beam, and the stress on the axis.

128. There are two equal rods connected by a smooth joint; the other extremity of one of the rods can move about a fixed point, and that of the second along a smooth horizontal axis passing through the fixed point, and about which the system is

revolving under the action of gravity. Find a differential equation to determine the inclination of the rods to the axis at any time.

129. An elliptic lamina whose excentricity is $\frac{1}{5}\sqrt{10}$ is supported with its plane vertical and transverse axis horizontal by two smooth, weightless pins passing through its foci. If one of the pins be suddenly released, show that the pressure on the other pin will be initially unaltered.

130. A plane lamina in the form of a circular sector whose angle is $2\,a$, is suspended from a horizontal axis through its centre, perpendicular to its plane. Find the time of a small oscillation, and show that if $3\,a = 4\sin a$ the time of oscillation will be the same about a horizontal axis through the extremity of the radius passing through the centre of inertia of the lamina.

131. A hollow cylinder open at both ends, of which the height is to the radius as 3 to $\sqrt{2}$, has a diameter of one of its ends fixed. Show that the centres of percussion lie on a straight line the distance of which from the fixed axis is eight-ninths of the height of the cylinder.

132. A lamina $ABCD$ is movable about AB as a fixed axis. Show that if CD be parallel to AB and $AB^2 = 3\,CD^2$, the centre of percussion will be at the intersection of AC and BD.

133. In the case of a rigid body freely rotating about a fixed axis, show that in order that a centre of percussion may exist the axis must be a principal axis with respect to some point in its length.

134. A uniform rod movable about one end, moves in such a manner as to make always nearly the same angle a with the vertical. Show that the time of its small oscillations is

$$2\pi\sqrt{\left\{\frac{2\,a\cos a}{3\,g(1+3\cos^2 a)}\right\}},$$

a being the length of the rod.

135. One end of a heavy uniform rod slides freely on a fine smooth wire in the form of an ellipse of excentricity $\frac{\sqrt{3}}{2}$, and axis minor equal to the length of the rod; the other end of the rod slides on a smooth wire coinciding with the axis minor of the ellipse. The system is set rotating about the latter wire, which is fixed in a vertical position. Prove that if θ be the inclination of the rod to the vertical at the time t, α the initial value of θ, and ω the initial angular velocity of the system about the vertical axis, $\cos \theta = \cos \alpha \cos (\omega t \sin \alpha)$.

136. A lamina in the form of an equilateral triangle rests with its base on a horizontal plane, and is capable of moving in a vertical plane about a hinge at one extremity of its base. Prove that it will turn completely over if it be struck at its vertex a blow greater than $2 mk \sqrt{\left(\dfrac{g}{a\sqrt{3}}\right)}$ in a direction perpendicular to that side which does not pass through the hinge, m being the mass, a the length of a side of the lamina, k its radius of gyration about an axis through one of its angular points perpendicular to its plane.

137. In the case of the motion of a rigid body about a horizontal axis under the action of gravity, show that the forces are reducible to a single force if the axis be a principal axis at the point where the perpendicular on it from the centre of gravity meets it and not otherwise. If the horizontal fixed axis be a principal axis at a point other than that at which the perpendicular on it from the centre of gravity meets it, and if the centre of gravity start from the horizontal plane passing through the fixed axis, determine the pressures.

138. An elliptic paraboloid, cut off by a plane parallel to the tangent plane at the vertex, is capable of freely rotating about a diameter of the base as a fixed axis. Find the line of action of an impulse which, acting on the paraboloid, produces no impulse on the fixed diameter.

139. If a rigid body have a centre of percussion with respect to a given axis, show that there is one with respect to any parallel axis, in a plane containing the given axis and the centre of inertia.

140. Investigate the angular velocity of the top (Fig. 51, page 117) while the string is still unwinding, assuming (1) the tension of the string to be constant and the axis to be cylindrical, (2) the increase of the tension of the string over the initial tension to vary as the length of string drawn off and the axis to be conical.

141. The part of a paraboloid of revolution cut off by a plane through the focus is fixed at a point in the circumference of its circular base. If it be struck by a blow at any point, in a direction parallel to its axis, find the initial instantaneous axis.

142. If but one force act on a rigid body, one point of which is fixed, the body's angular velocity about the instantaneous axis will be a maximum or a minimum when the instantaneous axis is perpendicular to the direction of the force.

143. A uniform rod can turn freely about one end which is fixed, the other end resting on a smooth inclined plane. If it be just disturbed from its position of unstable equilibrium, prove that it will never leave the plane unless its inclination to the horizon be $> \tan^{-1}(\frac{1}{5}\tan B)$, where B is the semi-vertical angle of the cone described by the rod.

144. A rigid body, fixed at one point only, is in motion under the action of finite forces. If, throughout the motion, the angular acceleration of the body about the instantaneous axis bear to the moment of inertia about this axis and to the forces acting on the body the same relation as if the axes were fixed, prove that if the three principal moments of inertia at the fixed point be not all equal the locus of the axis relatively to the body is a cone of the second order.

145. A triangular lamina ABC has the angular point C fixed, and is capable of free motion about it. A blow is struck at B,

perpendicular to the plane of the lamina. Show that the instantaneous axis passes through one of the points of trisection of the side AB.

146. Two equal uniform rods are capable of motion about a common extremity which is fixed, their upper ends being joined by an elastic string. They are set in vibration about a vertical axis bisecting the angle between them. If in the position of steady motion the natural length $(2\,l)$ of the string be doubled, the modulus of elasticity being equal to the weight of either rod, then the angular velocity about the vertical will be

$$\sqrt{\left\{\frac{3\,g(h-l)}{2\,hl}\right\}},$$

where h is the height of the string above the fixed extremity.

147. A rigid body, of which two of the principal moments at the centre of inertia are equal, rotates about a third principal axis, but this axis is constrained to describe uniformly a fixed right circular cone of which the centre of inertia is the vertex. Prove that the resultant angular velocity of the body is constant, that the requisite constraining couple is of constant magnitude, and that the plane of the couple turns uniformly in the body about the axis of unequal moment.

148. An ellipsoid is rotating with its centre fixed about one of its principal axes (that of x) and receives a normal blow at a point (h, k, l). If the initial axis of rotation after the blow lie in the principal plane of yz, its equation is

$$c^2(a^2 + c^2)(a^2 - b^2)ky + b^2(a^2 + b^2)(a^2 - c^2)lz = 0.$$

149. A sphere whose centre is fixed has an elastic string attached to one point, the other end of the string being fastened to a fixed point. To the sphere is given an angular velocity about an axis. Give the equations for determining its motion, the string being supposed stretched and no part of it in contact with the surface of the sphere. If the natural length of the

string be equal to a, the radius of the sphere, and it be fixed at a point O at a distance $= a(\sqrt{2} - 1)$ from its centre, and if the sphere be turned so that the point on it to which the string is fastened may be at the opposite extremity of the diameter through O, prove that the time of a complete revolution

$$= \sqrt{\left(\frac{a}{5\,ng}\right)\left(\pi + \frac{2\sqrt{2}}{3}\right)},$$

where $n = \dfrac{\text{modulus of elasticity}}{\text{weight of sphere}} = \sqrt{\left(\dfrac{aM}{5\,\mu}\right)\left(\pi + \dfrac{2\sqrt{2}}{3}\right)},$

where $\mu = $ modulus of elasticity.

150. If the angular velocities of a rigid body, at any time t, about the axes x, y, z, are proportional respectively to

$$\cot(m - n)t, \quad \cot(n - l)t, \quad \cot(l - m)t,$$

determine the locus of the instantaneous axis.

151. A uniform rod of length $2\,a$ can turn freely about one extremity. In its initial position it makes an angle of 90° with the vertical and is projected horizontally with an angular velocity ω. Show that the least angle it makes with the vertical is given by the equation $4\,a\omega^2 \cos\theta = 3\,g \sin^2\theta$.

152. A rigid body rotates about a fixed point under the action of no forces. Investigate the following equations, the invariable line being taken as the axis of z:

$$\frac{d\theta}{dt} = -G \sin\theta \sin\phi \cos\phi \left(\frac{1}{A} - \frac{1}{B}\right);$$

$$\frac{d\psi}{dt} = G\left(\frac{\cos^2\phi}{A} + \frac{\sin^2\phi}{B}\right);$$

$$\frac{d\phi}{dt} + \cos\theta \frac{d\psi}{dt} = \frac{G \cos\theta}{C};$$

G denoting the angular momentum of the body, and the other symbols having their usual meaning.

153. One point of a rigid body is fixed and the body is set in motion in any manner and left to itself under the action of no force. Prove that if A, B, C be its principal moments of inertia at the fixed point, G its angular momentum, λ its component angular velocity about the invariable line, ω its whole angular velocity, the component angular velocity of the instantaneous axis about the invariable line will be

$$\lambda + \frac{(G - A\lambda)(G - B\lambda)(G - C\lambda)}{ABC(\omega^2 - \lambda^2)}.$$

154. A rod is fixed at one end to a point in a horizontal plane about which it can move easily in any direction. When it is inclined to the horizon at a given angle, a given horizontal velocity is communicated to its other end. What will be the velocity and direction of the motion of the free end at the moment when the rod falls on the horizontal plane?

155. AD, BC are two equal rigid rods movable about a pin at L, such that $AL = DL = BC = CL$, and their ends are connected by four elastic strings of equal lengths. If the beams are made to revolve in opposite directions about L through a given angle, and then the system be left to itself, determine its subsequent motion.

156. AB, BC, CD are three equal beams connected by pins at B and C and lying in the same right line. If a given impulse be communicated to BC at its centre in a direction perpendicular to its length, determine the impulse on the pins.

157. A uniform rod is free to rotate about its extremity in a vertical plane, while that plane is constrained to revolve uniformly about a vertical axis through the extremity of the rod. Show that if the rod be let fall from an inclination of 30° above the horizon, it will just descend to the vertical position if $a\omega^2 = 3g$, where ω is the angular velocity of the plane and $2a$ is the length of the rod. Also explain the nature of the motion according as $a\omega^2$ is less than $3g$ or greater than $3g$.

158. A rigid rod of given mass can revolve about its middle point in a plane inclined at a given angle to the horizon. A given angular velocity is communicated to both rod and plane about a vertical axis through the middle point of the rod, the system being then left to itself. Show that the rod will oscillate about its horizontal position.

159. One of the principal axes of a body revolves uniformly in a fixed plane, while the body rotates uniformly about it. Determine the constraining couple and show that if the moments of inertia about the other two principal axes are equal, the couple has a constant moment.

160. A body, two of whose principal moments are equal, is free to rotate about its centre of gravity, which is fixed relatively to the earth's surface. Prove that if the body be made to rotate very rapidly about its principal axis of unequal moment, that axis will move both in altitude and azimuth, and that if the motion in altitude be prevented and the axis be originally placed horizontally in the meridian, it will be in a position of equilibrium, stable or unstable, according as the rotation is from west to east, or from east to west. If the axis be originally directed in any other azimuth, it will oscillate about its position of stable equilibrium nearly in the same way as the simple circular pendulum whose length $= Bg/(A\Omega\omega \cos \lambda)$, where A and B are the principal moments, Ω the angular velocity of the earth about its axis, ω that of the disc, and λ the latitude of the place of experiment.

161. A body turning about a fixed point of it is acted on by forces which always tend to produce rotation about an axis at right angles to the instantaneous axis. Show that the angular velocity cannot be uniform unless

$$\frac{C-B}{A} + \frac{B-A}{C} + \frac{A-C}{B} = 0,$$

A, B, C being the principal moments of inertia with respect to the fixed point.

162. If forces act on a homogeneous spheroid tending always to produce rotation about an axis α, in the plane of the equator, the instantaneous axis will describe a circular cone in the body about its polar axis; but the angular velocity about the instantaneous axis will not be uniform unless the axis α be always at right angles to the instantaneous axis.

163. A sphere movable about a point in its surface, which is fixed relatively to the earth, is in equilibrium under the action of gravity. Suppose the earth to suddenly cease rotating about its axis, find the instantaneous axis of rotation of the sphere and show that the angular velocity about it would be

$$\omega \cos \theta \sqrt{\left\{ 1 + \left(\frac{2+5\mu}{7}\right)^2 \tan^2 \theta \right\}},$$

ω being the angular velocity of the earth, μ the ratio between its radius and that of the sphere, and θ the latitude of the place.

164. A rigid body under the action of given forces is in motion about a fixed point. Defining the momental plane at any instant as that which would be the invariable plane if the forces affecting the body were at that instant to cease acting, show that if the body be constantly acted upon by a couple whose plane passes through the instantaneous axis and is normal to the momental plane, the distance of the momental plane from a fixed point will remain unchanged. If the body be acted upon at any instant by an impulsive couple in the plane referred to, show that the tangent of the angle through which the momental plane is suddenly turned varies as the moment of the couple.

165. A body is moving about a fixed point at a distance P from the invariable plane. Assuming that the central ellipsoid rolls upon the invariable plane, show that the equation to the surface generated by the instantaneous axis in the body is

$$Ax^2 + By^2 + Cz^2 = P^2(A^2x^2 + B^2y^2 + C^2z^2),$$

the equation to the central ellipsoid being $Ax^2 + By^2 + Cz^2 = 1$.

166. If the motion of a rigid body about a fixed point in it be represented by three coexistent angular velocities ω_x, ω_y, ω_z, about three axes mutually at right angles, show that all the particles in a cylindrical surface whose axis is $\dfrac{x}{\omega_x} = \dfrac{y}{\omega_y} = \dfrac{z}{\omega_z}$ will have linear velocities of equal magnitude. (See Prob. 2. p. 111.)

167. An equilateral triangular lamina is revolving in its own plane about its centre of inertia. If one of the angular points becomes suddenly fixed, show that the lamina will rotate about it with one-fifth of the original angular velocity.

168. A rigid body is free to move about a fixed point, and in the notation of Art. 62,

$$\omega_1 = a \sin \theta \sin \phi, \quad \omega_2 = a \sin \theta \cos \phi, \quad \omega_3 = a \cos \theta,$$

find the position of the body at any given time.

169. Show from Euler's Equations of Motion (Art. 60), that when no impressed forces act, no axis other than a principal axis can be a permanent axis.

170. When a body is acted on by no forces and moves about a fixed point, show that the locus of the instantaneous axis is a conical surface.

171. A prolate spheroid of revolution is fixed at its focus; a blow is given it at the extremity of the axis minor in a line tangent to the direction perpendicular to the axis major. Find the axis about which the body begins to rotate.

172. A rigid body fixed at a given point is free to rotate in any way about that point. Given the angular velocities about three axes mutually at right angles and fixed in space, find the velocity of any point in the body and the *vis viva* of the whole system.

173. In the case of a rigid body moving about a fixed point and subject to the action of no forces if the moment C be a har-

monic mean between the moments A and B, and the instantaneous axes describe the separating polhode, then will ϕ be constant, ψ will increase uniformly, and $\tan \theta = c^a \tan \theta_0$, where

$$c = \tfrac{1}{2} k^2 \left(\frac{1}{A} - \frac{1}{B} \right).$$

174. Integrate Euler's equations determining the motion of a rigid body about a fixed point for the case in which no forces act and two of the principal moments are equal.

175. If a body be in motion about a fixed point under the action of no external forces, show that the angular velocity about the radius vector of the momental ellipsoid, about which the body is turning, varies as that radius vector, and that the perpendicular on the tangent plane at the extremity of the radius vector is constant.

176. A plane lamina of uniform density and thickness, bounded by a curve represented by the equation $r = a + b \sin^2 2\theta$, moves about its pole as a fixed point. Show that if the lamina be under the action of no forces, its angular velocity will be constant, and its axis will describe a right cone in space.

177. A lamina in the form of a quadrant of a circle is fixed at one extremity of its arc and is struck a blow perpendicular to its plane at the other extremity. Find the velocities generated and the pressures on the fixed point. If θ be the inclination of the instantaneous axis to the radius vector through the fixed point, show that

$$\tan \theta = \frac{10 - 3\pi}{15\pi - 10}.$$

178. The point O of a rigid body is fixed in space, but the body is capable of free motion about the point. OA, OB, OC are the principal axes and A', B', C' are the principal moments of inertia of the body at O. Show that the couple necessary to keep the body moving so that OC shall describe a cone with

semi-vertical angle a uniformly about the fixed line OZ and COA shall maintain a constant inclination to ZOC, must be in the plane $x(C' - B')\cos \beta \cos a + y(C' - A')\sin \beta \cos a + z(A' - B')$ $\sin \beta \sin a = 0$, referred to OA, OB, OC as axes.

179. If the component angular velocities of a rigid body about a system of axes fixed in space be ω_x, ω_y, ω_z, and those about a system fixed in the body be ω_1, ω_2, ω_3, and if these coincide respectively with the former at the time t, prove that

$$\frac{d^2\omega_1}{dt^2} = \frac{d^2\omega_x}{dt^2} - \omega_y \frac{d\omega_z}{dt} + \omega_z \frac{d\omega_y}{dt}.$$

Examine this and get the equation for $\frac{d^2\omega_1}{dt^2}$ in terms of ω_x, ω_y, ω_z.

180. A body acted on by no forces, and having one point fixed, is such that if A, B, C are the principal moments of inertia at the fixed point, C is a harmonic mean between A and B. Show that if θ be the angle which the axis of C makes with the invariable line, and ϕ the angle which the plane of CA makes with the plane through the invariable line and the axis of C, then will $\sin^2 \theta \cos 2\phi$ be constant.

181. A rigid lamina, not acted on by any forces, has one point in it which is fixed, but about which it can turn freely. If the lamina be set in motion about a line in its own plane, the moment of inertia about which is Q, show that the ratio of its greatest to its least angular velocity is $A + Q : B + Q$, where A and B are the principal moments of inertia about axes in the plane of the lamina. If the lamina in the previous problem be bounded by an equiangular spiral and the intercept of the radius vector to the extremity of the curve, and if the fixed point be the pole,

$$A + Q : B + Q :: 1 + \cos 2\gamma \sin^2(\gamma - \beta) : 1 - \cos 2\gamma \cos^2(\gamma - \beta),$$

where the extreme radius vector is inclined to one principal axis at an angle γ and to the initial position of the instantaneous axis at an angle β.

P

182. A smooth ball of radius a moves around the circumference of a disc of radius $r+a$ and of four times the mass of the ball; the disc is supported at its centre and provided with a rim (whose weight may be neglected) sufficient to keep the ball from falling off. Show that the velocity of the ball, in order that the disc may maintain a constant inclination of 45° to the horizontal, is

$$\sqrt{\left\{\frac{g(r-a)^3}{2\sqrt{2}\,a(r+a)}\right\}}.$$

183. A rigid lamina in the form of a loop of a lemniscate $r^2 = a^2 \cos 2\,\theta$, not acted on by any force, is started with a given angular velocity about one of the tangent lines through its nodal point, the nodal point being fixed. Prove that its greatest angular velocity has to its least angular velocity the ratio

$$\sqrt{(3\,\pi+4)} : \sqrt{(3\,\pi)}.$$

184. A rigid body, movable about a fixed point, is struck a blow of given magnitude at a given point. If the angular velocity thus impressed upon the body be the greatest possible, prove that

$$\frac{a}{l}\left(\frac{1}{B^2}-\frac{1}{C^2}\right)+\frac{b}{m}\left(\frac{1}{C^2}-\frac{1}{A^2}\right)+\frac{c}{n}\left(\frac{1}{A^2}-\frac{1}{B^2}\right)=0,$$

where A, B, C are the moments of inertia of the body about the principal axes at the fixed point, a, b, c are the coördinates of the point struck in relation to the principal axes at the fixed point, and l, m, n are the direction-cosines of the line of action of the blow.

185. A square lamina with one angle attached to a fixed point rotates about a side. What must be the angular velocity of the lamina in order that the side about which it rotates may remain vertical?

186. A rigid body is rotating about an axis through its centre of inertia, when a certain point of the body becomes suddenly fixed, the axis being simultaneously set free. Prove that if the

new instantaneous axis be parallel to the original fixed axis, the point must lie in the line represented by the equations

$$a^2lx + b^2my + c^2mz = 0, \quad (b^2 - c^2)\frac{x}{l} + (c^2 - a^2)\frac{y}{m} + (a^2 - b^2)\frac{z}{n} = 0,$$

the principal axes through the centre of inertia being taken as axes of coördinates, a, b, c the radii of gyration about these lines, and l, m, n the direction-cosines of the originally fixed axis referred to them.

187. An elliptic lamina, fixed at the focus, is struck in a direction perpendicular to its plane. Find the instantaneous axis and show that if the blow be applied at any point of the ellipse

$$\frac{y^2}{(1 - e^2)^2} + \frac{x^2}{(1 + 4e^2)^2} = c^2$$

the angular velocity will be the same, the focus being origin, and the axis major and latus rectum the axes of x and y respectively, and e being the excentricity.

188. A uniform rod of length a, freely movable about one end, is initially projected in a horizontal plane with angular velocity ω about the fixed point. If θ be the angle which the rod makes with the vertical and ϕ be the angle which the projection of the rod on the horizontal plane makes with the initial position, show that the equations of motion are

$$\sin^2\theta\,\frac{d\phi}{dt} = \omega, \quad \left(\frac{d\theta}{dt}\right)^2 = \frac{3g}{a}\cos\theta - \omega^2\cot^2\theta.$$

Find the lowest position of the rod and if this be when $\theta = \frac{\pi}{3}$, show that the resolved vertical pressure on the fixed point is then equal to $3\frac{1}{16}$ of the weight of the rod.

189. A lamina having one point fixed is at rest and is struck a blow perpendicular to its plane at a point whose coördinates, referred to the principal axes at the fixed point, are a, b. Show that the equation to the instantaneous axis is $ah^2x + bk^2y = 0$, h, k being the radii of gyration about the principal axes. Show

that if ab lie on a certain straight line, there will be no impulse
on the fixed point.

190. A uniform rod of length $2a$ and mass m, capable of free
rotation about one end, is held in a horizontal position, and on
it is placed a smooth particle of mass p at a distance c from the
point, c being $< \dfrac{4a}{3}$; the rod is then let go. Find the initial pressure of the particle on the rod, and show that the radius of
curvature of the particle's path is

$$\frac{9c^2}{4a-3c}\left(1+\frac{cp}{am}\right).$$

191. A lamina in the form of a symmetrical portion of the
curve $r = a(n\pi^2 - \theta^2)$ is placed on a smooth plane with its axis
vertical, then infinitesimally displaced and allowed to fall in its
own plane. If the lamina be loaded so that its centre of inertia is
at the pole and its radius of gyration $= 2a$, find the time in which
its axis will fall from one given angular position to another.

192. An elliptical lamina stands on a perfectly rough inclined
plane. Find the condition that its equilibrium may be stable,
and determine the time of a small oscillation.

193. A perfectly rough plane, inclined at a fixed angle to the
vertical, rotates about the vertical with uniform angular velocity.
Show that the path of a sphere placed upon the plane is given
by two linear differential equations of the form,

$$\frac{d^2y}{dt^2}+A\frac{dx}{dt}+By=0, \quad \frac{d^2x}{dt^2}-A'\frac{dy}{dt}+B^1x=C,$$

the origin being the point where the vertical line, about which
the plane revolves, meets the plane; the axis of y being the
straight line in the plane which is always horizontal.

194. The equal uniform beams AB, BC, CD, DE, are connected by smooth hinges and placed at rest on a smooth horizontal plane, each beam at right angles to the two adjacent,
so as to form a figure resembling a set of steps. An impulse

is given at the end A, along AB; determine the impulsive action on any hinge.

195. A rectangle is formed of four uniform rods of lengths $2a$ and $2b$ respectively, which are connected by smooth hinges at their ends. The rectangle is revolving about its centre on a smooth horizontal plane with an angular velocity ω, when a point, in one of the sides of length $2a$, suddenly becomes fixed. Show that the angular velocity of the sides of length $2b$ immediately becomes $\dfrac{3a+b}{6a+4b}\omega$. Find, also, the change in the angular velocity of the other sides and the impulse at the point which becomes fixed.

196. A uniform revolving rod, the centre of inertia of which is initially at rest, moves in a plane under the action of a constant force in the direction of its length. Prove that the square of the radius of curvature of the path of the rod's centre of inertia varies as the versed sine of the angle through which the rod has revolved at the end of any time from the beginning of the motion.

197. Six equal uniform rods are freely joined together and are at rest in the form of a regular hexagon on a smooth horizontal plane. One of the rods receives an impulse at its midpoint, perpendicularly to its length, and in the plane of the hexagon. Prove that the initial velocity of the rod struck is ten times that of the rod opposite to it.

198. A uniform rod of length $2a$ lies on a rough horizontal plane, and a force is applied to it in that plane and perpendicularly to its length at a distance P from its midpoint, the force being the smallest that will move the rod. Show that the rod begins to turn about a point distant $\sqrt{(a^2+p^2)}-p$ from the midpoint.

199. AB is a rod whose end A is fixed and which has an equal rod BC attached at B. Initially the rods AB, BC are in the same straight line, AB being at rest and BC on a smooth

horizontal plane having an angular velocity ω. Show that the greatest angle between the rods at any subsequent time is $\cos^{-1}\frac{5}{12}$ and that when they are again in a straight line, their angular velocities are $\frac{5\omega}{8}$ and $-\frac{3\omega}{8}$ respectively.

200. A rectangular board moving uniformly without rotation in a direction parallel to one side, on a smooth horizontal plane, comes in contact with a smooth fixed obstacle. Determine at what point the impact should take place in order that the angular velocity generated may be a minimum.

201. Four equal uniform rods, freely jointed at their extremities, are lying in the form of a square on a smooth horizontal table, when a blow is applied at one of the angles in a direction bisecting the angle. Find the initial state of motion of each rod, and prove that during the subsequent motion the angular velocity will be uniform.

202. A sphere is moving at a given moment on an imperfectly rough horizontal table with a velocity v, and at the same time has an angular velocity ω round a horizontal diameter, the angle between the direction of v and the axis of ω being u. Prove that the centre of the sphere will describe a parabola if

$$ak^2\omega^2 + (a^2 - k^2)v\omega \sin u = av^2.$$

203. Two rods, OA and OB, are fixed in the same vertical plane, with the point O upwards, the rods being at the same angle u to the vertical. The ends of a rod AB of length $2a$ slide on them. Show that if the centre of inertia of AB be its middle point, and the radius of gyration about it be k, the time of a complete small oscillation is

$$2\pi\sqrt{\left\{\frac{a^2\tan^2 u + k^2}{ag\cot u}\right\}}.$$

204. One end of a heavy rod rests on a horizontal plane and against the foot of a vertical wall; the other end rests against a parallel vertical wall, all the surfaces being smooth. Show that, if the rod slip down, the angle ϕ, through which it will turn

round the common normal to the vertical walls, will be given by the equation $\dfrac{d^2\phi}{dt^2}(1 + 3\cos^2\phi) + \dfrac{6g}{\sqrt{(a^2 - b^2)}}\sin\phi = C$, where $2a$ is the length of the rod and $2b$ the distance between the walls.

205. Two equal uniform rods, loosely jointed together, are at rest in one line on a smooth horizontal table, when one of them receives a horizontal blow at a given point. Determine the initial circumstance of the motion, and prove that, when next the rods are in a straight line, they will have interchanged angular velocities.

206. One end of a uniform rod of weight w can slide by a smooth ring on a vertical rod, the other end sliding on a smooth horizontal plane. The rod descends from a position inclined at an angle β to the horizon. Show that the rod will not leave the horizontal plane during the descent, but that its maximum pressure against it is $\frac{1}{4}w\cos^2\beta$ and that its ultimate pressure is $\frac{1}{4}w$.

207. A lamina capable of free rotation about a given point in its own plane, which point is fixed in space, moves under the action of given forces. If the initial axis of rotation of the lamina coincide very nearly with the axis of greatest moment of inertia in the plane of the lamina, the angular velocities about the other principal axes will be in a constant ratio during the motion.

208. A sphere of radius a is partly rolling and partly sliding on a rough horizontal plane. Show that the angle the direction of friction makes with the axis of x is $\tan^{-1}\dfrac{u + a\omega_1}{v - a\omega_2}$, u and v being the initial velocities, ω_1, ω_2 the initial angular velocities.

209. A perfectly rough circular cylinder is fixed with its axis horizontal. A sphere is placed on it in a position of unstable equilibrium, and projected with a given velocity parallel to the axis of the cylinder. If the sphere be slightly disturbed in a

horizontal direction perpendicular to the direction of the axis of the cylinder, determine at what point it will leave the cylinder.

210. A parabolic lamina, cut off by a chord perpendicular to its axis, is kept at rest in a horizontal position by three vertical strings fastened to the vertex and two extremities of the chord; if the string which is fastened to the vertex be cut, the tension of the others is suddenly decreased one-half.

211. Three equal, perfectly rough, inelastic spheres are in contact on a horizontal plane; a fourth equal sphere, which is rotating about its vertical diameter, drops from a given height and impinges on them simultaneously. Investigate the subsequent motion.

212. A rod of length a, moving with a velocity v perpendicular to its length on a smooth horizontal plane, impinges on an inelastic obstacle at a distance c from its centre. Show that its angular velocity when the end quits the obstacle is $\dfrac{3\sqrt{c}}{a^2}$.

213. A solid regular tetrahedron is placed with one edge on a smooth horizontal table and is allowed to fall from its position of unstable equilibrium. Find the angular velocity of the tetrahedron just before a face of it reaches the table, and the magnitude of the resultant impulsive blow.

214. A uniform sphere of radius a, when placed upon two parallel, imperfectly rough, horizontal bars, has its centre at a height b above the horizontal plane which contains the bars. It is started with a velocity v parallel to the bars, and an angular velocity ω about a horizontal axis perpendicular to the bars in such a direction as to be diminished by friction. In the case in which $2a^2\Omega > 5bv$, the sphere will begin to roll after a time

$$\frac{2ab(v + b\Omega)}{\mu g(2a^2 + 5b^2)},$$

where μ is the coefficient of friction. What will at that instant be the velocity and position of the sphere?

215. A heavy uniform rod slips down with its extremities in contact with a smooth horizontal floor and a smooth vertical wall, not being initially in a plane perpendicular to both wall and floor. Prove that if θ be the inclination to the horizon and ϕ the angle which the projection of the rod on the floor makes with the normal to the wall,

$$(k^2 + a^2)\sin\phi\frac{d^2(\cos\theta\cos\phi)}{dt^2} = k^2\cos\phi\frac{d^2(\cos\theta\sin\phi)}{dt^2},$$

and

$$(k^2 + a^2)\cos\theta\sin\phi\frac{d^2(\sin\theta)}{dt^2}$$
$$= k^2\sin\theta\frac{d^2(\cos\theta\sin\phi)}{dt^2} - ag\cos\theta\sin\phi,$$

$2a$ being the length of the rod and k its radius of gyration about an axis perpendicular to it through the centre of inertia.

216. A body possesses given motions of translation and rotation referred to a given point of it. Find under what condition the motion may be exhibited by rotation about a single axis, and the equations to this axis when the condition is satisfied.

217. A heavy straight rod slides freely over a smooth peg. Show that the equations to its motion are

$$\frac{d^2r}{dt^2} - r\left(\frac{d\theta}{dt}\right)^2 = g\sin\theta,$$

and

$$\frac{d}{dt}\left\{(r^2 + k^2)\frac{d\theta}{dt}\right\} = gr\cos\theta,$$

where r and θ are coördinates of the centre of inertia reckoned from the peg and a horizontal line.

218. A smooth wire of given mass is bent into the form of an ellipse and laid upon a smooth horizontal table; an insect of given weight is gently laid on the wire and crawls along it. Find the path described by the centre of the elliptic wire and trace it on the table.

219. The effect of an earthquake being assumed to be a sudden horizontal displacement in a given direction of every body

fixed to the surface of the earth, explain the nature of the motion caused by the shock in the half of a uniform cylindrical stone column which is cut off by a plane bisecting the cylinder diagonally, and which rests with its base upon a fixed horizontal plane, friction being supposed the same at every point.

220. If a rigid body initially at rest be acted on by given impulses, whose resultant is a single impulse, show that the axis of instantaneous rotation will be perpendicular to the direction of that resultant.

221. A circular disc rolls down a rough curve in a vertical plane. If the initial and final positions of the centre of the disc be given, show that when the time of motion is the least possible the curve on which the disc rolls is an involute of a cycloid.

222. A circular ring is free to move on a smooth horizontal plane on which it lies, and an uniform rod has its extremities connected with and movable on the smooth arc of the ring. The system being set in motion on the plane, show that the angular velocity of the rod is constant, and describe the paths of the centres of the rod and ring.

223. A wheel whose centre of gravity does not coincide with the centre of the figure is allowed to roll down an inclined plane which is so rough as to prevent sliding. If α be the inclination of the plane, a the radius of the wheel, h the distance of its centre of inertia from the centre of the figure, and k the radius of gyration of the wheel about an axis through its centre of inertia perpendicular to its plane, show that when the wheel has rolled from rest through an angle γ, the resistance exerted by the plane either equals zero or is normal to the plane, γ being given by the equation,

$$[\tan \alpha \tan \tfrac{1}{2} \gamma \{(a+h)^2+k^2-a^2\} +a^2]^2$$
$$= a^4 - \{(a-h)^2+k^2+a^2\} \{(a+h)^2+k^2-a^2\} \tan^2 \alpha.$$

224. A sphere on a smooth horizontal plane is placed in contact with a rough vertical plane which is made to revolve with a uniform angular velocity ω about a vertical axis in itself. If a be the initial distance of the point of contact from the axis, r the distance after a time t, and c the radius of the sphere, prove that $2r = (a + c\sqrt{\tfrac{7}{5}})e^{\omega t \sqrt{\frac{7}{5}}} + (a - c\sqrt{\tfrac{7}{5}})e^{-\omega t \sqrt{\frac{7}{5}}}$. Also show that as t increases indefinitely, the ratio of the friction to the pressure approximates to $1 : \sqrt{35}$.

225. A free plane lamina receives a single blow perpendicular to its plane. Show that (i) if the locus of points where the blow may have been applied be a straight line, the spontaneous axis will pass through a determinate point, (ii) if the locus be a circle (centre C), the spontaneous axis will be a tangent to an ellipse whose axes are in the direction of the principal axes at C in the plane of the lamina.

226. A sphere, in contact with two fixed rough planes, rolls down under the action of gravity. If 2α be the angle between the planes which are equally inclined to the horizon, and with which their line of intersection makes an angle β, show that the acceleration of the centre of the sphere is uniform and equal to

$$\frac{5 \sin^2 \alpha \sin \beta}{2 + 5 \sin^2 \alpha} g.$$

227. Three equal smooth spheres are placed in contact, each with the other two, on a smooth horizontal plane, and connected at the points of contact. A fourth equal sphere is then placed so as to be supported by the other three. Supposing the connections between the three spheres suddenly destroyed, show that the pressure between the fourth sphere and each of the other three is suddenly diminished by one-seventh. Also determine the subsequent motion.

228. A sphere is placed upon two smooth equal spheres held in contact, and these rest on a smooth horizontal plane in the position of equilibrium. Show if the spheres be left to them-

selves, the pressure on the upper sphere is instantaneously diminished to six-sevenths of its former amount.

229. A plane lamina lies on a smooth horizontal table. If one point of it be constrained to move uniformly along a straight line on the table, show that the lamina will revolve about the point with uniform angular velocity, and determine the magnitude and direction of the force of constraint at any time.

230. A sphere has an angular velocity about a horizontal diameter and falls upon a rough, inelastic board which is moving uniformly in a horizontal plane in the direction of this diameter. Find the initial direction of the motion and its path afterwards.

231. If the velocities of two given points of a rigid body be given in magnitude and direction, determine the velocity of any other point in the body.

232. Prove that any motion of a rigid rod may be represented by a single rotation about any one of an infinite number of axes, and find the locus of these axes.

233. A free ellipsoid is struck a blow normal to its surface. Show that, in general, there is no axis of spontaneous rotation.

234. A free rigid body is at a certain moment in a state of rotation about an axis through its centre of inertia, when another point in the body suddenly becomes fixed. Prove that there are three directions of the original instantaneous axis for which the new instantaneous axis will be parallel to it, and that these directions are along conjugate diameters of the momental ellipsoid at the centre of inertia.

235. A little squirrel clings to a thin rough hoop, of which the plane is vertical and is rolling along a perfectly rough horizontal plane. The squirrel makes a point of keeping a constant altitude above the horizontal plane and selects his place on the hoop so as to travel from a position of instantaneous rest, the greatest possible distance in a given time. Prove that *m*

being the weight of the squirrel and m' that of the hoop, the inclination of the squirrel's distance from the centre of the hoop to the vertical is equal to $\cos^{-1}\left(\dfrac{m}{m+2\,m'}\right)$.

236. A rough homogeneous sphere rests on a rough horizontal plane; a heavy inelastic beam sliding through two smooth rings in the same vertical line falls upon it from a given height. Find the position of the sphere relatively to the beam, in order that the angular velocity communicated to the sphere may be the greatest possible.

237. Two equal uniform rods, freely jointed together at one extremity of each, are at rest on a smooth horizontal plane. Find the point at which either must be struck in order that the system may begin to move as if it were rigid.

238. A heavy beam is placed with one end on a smooth inclined plane and is left to the action of gravity. If the vertical plane constraining the beam be perpendicular to the inclined plane, find the motion of the beam and the pressure on the plane when a given angle has been turned through.

239. A disc rolls upon a straight line on a horizontal plane, the disc moving with its flat surface in contact with the plane. Show that the disc will be brought to rest after a time $\dfrac{27\,\pi v}{64\,\mu}$ where v is the initial velocity of the centre, and μ the coefficient of friction between the disc and the table.

240. Determine how a free rigid body at rest must be struck in order that it may rotate about a fixed axis.

241. A uniform bar is constrained to move with its extremities on two fixed rods at right angles to each other, and is under the action of an attraction varying as the distance from, and tending to, the point of intersection of the rods. Determine the time of a small oscillation when the bar is slightly displaced from the position of rest.

242. A heavy cycloid, the radius of whose generating circle is $\frac{a}{4}$, is mounted so as to admit of sliding in a vertical plane with its base always horizontal and so that every point of it moves in a straight line, inclined at an angle of 45° to the horizontal. A uniform, smooth, heavy chain of length a and mass equal to that of the cycloid is laid over it so as to be in equilibrium when the cycloid is supported; if the support be suddenly removed, find the tension at any point at the commencement of motion and show that it is a maximum at a distance from the vertex given by the equation

$$8 \pi a = (96 - \pi^2)s - 32\sqrt{(a^2 - s^2)}.$$

243. A body is turning about an axis through its centre of inertia; a point in the body suddenly becomes fixed. If the new instantaneous axis be a principal axis with respect to the point, show that the locus of the point is a rectangular hyperbola.

244. A uniform rod of mass m and length $2a$ has attached to it a particle of mass p by a string of length b. The rod and string are placed in a straight line on a smooth horizontal plane, and the particle is projected with velocity v at right angles to the string. Prove that the greatest angle which the string makes with the rod is

$$2 \sin^{-1} \frac{\sqrt{\left\{ a\left(1 + \frac{m}{p}\right) \right\}}}{12\,b}$$

and that the angular velocity at the instant is $\dfrac{v}{a+b}$.

245. A rough sphere is projected on a rough horizontal plane and moves under an acceleration tending to a point in the plane and varying as the distance from that point. Show that the centre of the sphere will describe an ellipse, and find its component angular velocities in terms of the time.

246. Three equal uniform rods, AB, BC, CD, freely jointed together at B and C, are lying in a straight line on a smooth

horizontal plane and a given impulse is applied at the midpoint of BC at right angles to BC. Determine the velocity of BC when each of the other rods makes an angle θ with it, and prove that the directions of the stresses at B and C make with BC angles equal to $\tan^{-1}(\frac{2}{3}\tan\theta)$.

247. Three equal uniform straight lines, AB, BC, CD, freely jointed together at B and C, are placed in a straight line on a smooth horizontal plane and one of the outside rods receives a given impulse in a direction perpendicular to its length at its midpoint. Compare the subsequent stresses on the hinges with the impulse given to the rod.

248. A homogeneous right circular cylinder of radius a, rotating with angular velocity ω about its axis, is placed with its axis horizontal on a rough inclined plane so that its rotation tends to move it up the plane. If α be the inclination of the plane to the horizontal and $\tan\alpha$ the coefficient of friction, show that the axis of the cylinder will remain stationary during a period $T = \dfrac{a\omega}{2g\sin\alpha}$ and that its angular velocity at any time t during this period is equal to $\omega - \dfrac{2gt\sin\alpha}{a}$.

249. A hoop is hung upon a horizontal cylinder of given radius. Determine the time of a small oscillation
    I. When the cylinder is rough.
    II. When the cylinder is smooth.

250. Prove the following equations for determining the motion of a rigid body whose principal moments of inertia at the centre of inertia are equal:

$$\frac{X}{G} = \frac{du}{dt} - v\theta_3 + w\theta_2, \text{ etc.,} \quad \frac{L}{A} = \frac{d\omega_1}{dt} - \omega_2\theta_3 + \omega_3\theta_2, \text{ etc.;}$$

u, v, w being the velocities of the centre of inertia parallel to the three axes moving in space, ω_1, ω_2, ω_3 the angular velocities about these axes, θ_1, θ_2, θ_3 the angular velocities of these axes about fixed axes instantaneously coincident with them, X, Y, Z

the resolved forces, L, M, N their moments about the axes, G the mass of the body, and A its moment of inertia about any axis through the centre of inertia.

251. A uniform rod of mass m and length $2a$ has attached to it a particle of mass p by means of a string of length l; the rod and string are placed in one straight line on a smooth horizontal plane, and the particle is projected with a velocity v at right angles to the string. Prove, then, when the rod and string make angles θ, ϕ with their initial positions,

$$\left\{ k^2 + ab\cos(\phi - \theta) \right\}\frac{d\theta}{dt} + \left\{ b^2 + ab\cos(\phi - \theta) \right\}\frac{d\phi}{dt} = (a + b)v,$$

$$k^2\left(\frac{d\theta}{dt}\right)^2 + 2ab\cos(\phi - \theta)\frac{d\theta}{dt}\cdot\frac{d\phi}{dt} + b^2\left(\frac{d\phi}{dt}\right)^2 = v^2,$$

where
$$k^2 = a^2\frac{4p + 3m}{3p + 3m}.$$

252. A sphere of radius a is projected on a rough horizontal plane so as partly to roll and partly to slide. If the initial velocity of translation be v, the initial rotation ω about a horizontal axis, and the direction of the former make an angle α with the axis of the latter, show that the angle through which the direction of motion of the centre has turned, when perfect rolling begins, is

$$\tan^{-1}\frac{2a\omega\cos\alpha}{5v - 2a\omega\sin\alpha}.$$

253. If a homogeneous sphere roll on a perfectly rough plane under the action of any forces whatever, of which the resultant passes through the centre of the sphere, the motion of the centre of inertia will be the same as if the plane were smooth and all the forces were reduced in a certain constant ratio; and the plane is the only surface which possesses this property.

254. A smooth ring of mass m slides on a uniform rod of mass M. Determine the velocity of the ring at any point of the rod which it reaches, no impressed forces being supposed to act.

If when the ring is distant c from the centre of the rod, the angle at which its path is inclined to the instantaneous position of the rod be greater than $\cot^{-1}\left\{2 + \dfrac{mc^2}{(M+m)k^2}\right\}$, show that it will never reach the centre of the rod, k^2 being the radius of gyration of the rod about its centre.

255. A uniform rod of weight W and length $2a$ is supported in a horizontal position by two fine vertical threads, each of length c, and each is attached at a distance c from the centre of the rod. The rod is slightly displaced by the action of a horizontal couple whose moment is bW, and which does not move the centre of the rod out of a vertical line. Show that the time of a small oscillation of the rod will be

$$4\pi\left\{\frac{(a^2 + 3\,b^2)}{3\,g\sqrt{(c^2 - b^2)}}\right\}^{\frac{1}{2}}.$$

256. A circular lamina, rotating about an axis through the centre perpendicular to its plane, is placed in an inclined position on a smooth horizontal plane. Give a general explanation of the motion deduced from dynamical principles, and show that under certain circumstances the lamina will never fall to the ground, but that its centre will perform vertical oscillations, the time of an oscillation being

$$\frac{\pi}{2}\left(\frac{1 + 4\cos^2\alpha}{\omega^2 - \dfrac{g}{a}\sin\alpha}\right)^{\frac{1}{2}},$$

α being the inclination of the lamina to the horizon at first, a its radius, and ω its angular velocity.

257. A beam rests with one end on a smooth horizontal plane, and has the other suspended from a point above the plane by a weightless, inextensible string; the beam is slightly displaced in the plane of beam and string. Find the time of a small oscillation.

Q

258. Find the condition that a free rigid body in motion may be reduced to rest by a single blow.

259. A perfectly rough horizontal plane is made to rotate with constant angular velocity about a vertical axis which meets the plane in O. A sphere is projected on the plane at a point P so that the centre of the sphere has initially the same velocity in direction and magnitude as if the sphere had been placed freely on the plane at a point Q. Show that the sphere's centre will describe a circle of radius OQ, and whose centre R is such that OR is parallel and equal to OP.

260. If a free rigid body be struck with a given impulse, and any point of the body be initially at rest after the blow, show that a line of points will also be at rest, and determine the condition that this may be the case in a body previously at rest.

261. A free rigid body of mass m is at rest, its moments of inertia about the principal axes through its centre of inertia being A, B, C. Supposing the body to be struck with an impulse R through its centre of inertia, and with an impulsive couple G, prove that it will revolve for an instant about an axis whose velocity is in the direction of its length and equal to

$$\frac{\dfrac{LX}{Am}+\dfrac{MY}{Bm}+\dfrac{NZ}{Cm}}{\left(\dfrac{L^2}{A^2}+\dfrac{M^2}{B^2}+\dfrac{N^2}{C^2}\right)^{\frac{1}{2}}},$$

X, Y, Z being the components of R, and L, M, N the components of G, in the principal planes.

262. A sphere with a sphere within it, the diameter of the latter being equal to the radius of the former, is placed on a perfectly rough inclined plane, with the centre of inertia at its shortest distance from the plane, and is then left to itself. Find the angular velocity of the body when it has rolled round just once, and determine the pressure then upon the plane.

263. Two equal rods of the same material are connected by a free joint and placed in one straight line on a smooth horizontal plane; one of them is struck perpendicularly to its length at its extremity remote from the other rod. Show that the linear velocity communicated to its centre of inertia is one-fourth greater than that which would have been communicated to it by a similar blow if the rod had been free.

In the subsequent motion show that the minimum angle which the rods make with one another is $\cos^{-1}\frac{1}{3}$.

264. AB, BC, CD are three equal uniform rods lying in a straight line on a smooth horizontal plane, and freely jointed at B and C; a blow is applied at the midpoint of BC. Show that if ω be the initial angular velocity of AB or CD, θ the angle which they make with BC at time t,

$$\frac{d\theta}{dt} = \frac{\omega}{\sqrt{(1 + \sin^2\theta)}}.$$

265. A lamina of any form lying on a smooth horizontal plane is struck a horizontal blow. Determine the point about which it will begin to turn, and prove that if c, c' be the distances from the centre of inertia of the lamina of this point and of the line of action of the blow respectively, $cc' = k^2$, where k is the radius of gyration of the lamina about the vertical line through its centre of inertia.

266. A circular lamina whose surface is rough is capable of revolution about a vertical axis through its centre perpendicular to its plane, and a particle whose mass is equal to that of the lamina is attached to the axis by an inelastic string and rests on the lamina. If the lamina be struck a blow in its own plane, determine the motion.

267. A bicycle whose wheels are equal and body horizontal is proceeding steadily along a level rough road. Obtain equations for determining the instantaneous impulses on the machine when the front wheel is suddenly turned through a horizontal angle θ.

Show that the initial horizontal angular velocity is proportional to the original velocity.

268. The radii of the portions of a horizontal differential axle of weight W are a and b, and their lengths are b and a. The suspended weight is also W. If the balancing power be removed and the weight be allowed to fall, show that in time it will fall through

$$\frac{(a-b)^2}{3(a-b)^2 + 2ab} g t^2.$$

269. Show how to determine the angular velocities of a rotating mass by observations of the instantaneous direction cosines of points on its surface referred to three fixed rectangular axes, and their time rates of increase; i.e. $\frac{dl}{dt}$, etc. How many such observations are necessary?

270. A sphere composed of an infinite number of infinitely thin concentric shells is rotating about a common axis under no forces. Assuming that the friction of any shell on the consecutive external one at any point varies as the square of the angular velocity and the distance of the point from the axis, obtain the equation $kr\frac{d\omega}{dt} - r\omega\frac{d\omega}{dr} = 2\omega^2$ for the angular velocity at any time of shell of radius r, and show that the solution of this equation is $r^2\omega = f\left(\frac{kr}{3\omega} + t\right)$, where f is an arbitrary function.

271. An egg with its axis horizontal is rolling steadily round a rough vertical cone of semi-vertical angle a. The shape, weight, moment of inertia, etc., of the egg being known, find the friction acting, and the time of completing a circuit.

272. A vertical, double, elastic, wire helix is rigidly attached at one end to a horizontal bar, mass M, and is constrained to retain the same radius a. When in equilibrium the tangent angle is a. An additional weight $Mg\theta$, or a torsion couple $Mga\theta$, can alter a into $a + \theta$. If the bar be depressed, and consequently turned

through an angle, show that the time of a small oscillation will be $2\pi\sqrt{\left\{\dfrac{l}{g}(\cos\alpha + \dfrac{n^2}{3}\sin\alpha)\right\}}$, where l and $2\,na$ are the lengths of the helix and the bar respectively.

273. Four equal, smooth, inelastic, circular discs of radius a are placed in one plane with their centres at the four corners of a square of which each side $= 2\,a$. They attract one another with a force varying as the distance. A blow is given to one of them in the line of one of the diagonals of the square. Investigate the whole of the subsequent motion.

274. P and Q are two points in a uniform rod equidistant from its centre. The rod can move freely about a hinge at P. The hinge is constrained to move up and down in a vertical line. If the motion be such that Q moves in a horizontal line, determine the velocity when the rod has any given inclination, the rod being supposed to start from rest in a horizontal position.

In the case in which the whole length of the rod $= PQ\sqrt{3}$, show that the time of a complete oscillation is $(2\pi)^{\frac{3}{2}}(\Gamma\frac{1}{4})^{-2}$.

275. A circular and a semicircular lamina of equal radii a are made of the same material, which is perfectly rough. Their centres are joined by a tight inelastic cord; also the centre of the circular lamina is joined to the highest point of the semicircular lamina by a string of length $a\sqrt{3}$. The semicircular lamina stands with its base on a perfectly rough, inelastic plane. The circular lamina rests on the top of the semicircular lamina and in the same vertical plane with it. It is disturbed from its position of equilibrium. Prove that just after it has struck the plane its angular velocity $= \dfrac{8}{13}\sqrt{\left(\dfrac{g}{3\,a}\right)}$.

276. A uniform rod, capable of free motion about one extremity, has a particle attached to it at the other extremity by means of a string of length l and the system is abandoned freely to the action of gravity when the rod is inclined at an angle α to

the horizon and the string is vertical. Prove that the radius of curvature of the particle's initial path is

$$9l \cdot \frac{m+2p}{m} \cdot \frac{\cos^3 \alpha}{\sin \alpha (2 - 3\sin^2 \alpha)},$$

m and p being the masses of the rod and particle respectively.

277. To a smooth horizontal plane is fastened a hoop of radius c, which is rough inside, μ being the coefficient of friction. In contact with this a disc of radius a is spun with initial angular velocity n and its centre is projected with velocity v in such a direction as to be most retarded by friction. Show that after a time $\dfrac{c-a}{\mu} \cdot \dfrac{k^2}{v} \cdot \dfrac{an+v}{a^2v-ank^2}$ the disc will roll on the inside of the hoop.

278. An elephant rolls a homogeneous sphere of diameter a inches and mass S directly up a perfectly rough plane inclined β to the horizon, by balancing himself at a point distant α from the sphere's highest point at each instant. Show that, the elephant being conceived as without magnitude but of mass E, he will move the sphere through a space

$$\frac{t^2}{2} \cdot \frac{g}{a} \cdot \frac{E \sin \alpha - (E+S)\sin \beta}{E \cos(\alpha+\beta) + E + \frac{7}{5}S},$$

where t is the time elapsed since the commencement of the motion.

279. A circular disc of mass M and radius r can move about a fixed point A in its circumference, and an endless fine string is wound round it carrying a particle of mass m, which is initially projected from the disc at the other end of the diameter through A, with a velocity v normally to the disc, which is then at rest. Show that the angular velocity of the string will vanish when the length of the string unwound is that which initially subtended at the point A an angle β given by the equation

$$(\beta \tan \beta + 1)\cos^2 \beta + \frac{3}{8}\frac{M}{m} = 0,$$

and that the angular velocity of the disc is then $\dfrac{v}{2r}(2 + \beta \tan \beta)^{-1}$.

280. A uniform rod AB can turn freely about the end A, which is fixed, the end B being attached to the point C, distant c vertically above A, by an elastic string which would be stretched to double its length by a tension equal to the weight of the rod. If the rod be in equilibrium when horizontal and be slightly displaced in a vertical plane, prove that the period of its small oscillations is $\dfrac{2\pi}{c}\sqrt{\left(\dfrac{\rho^3}{3g}\right)}$, where ρ is the stretched length of the string in equilibrium.

281. A hollow cylinder, of which the exterior and interior radii are a and b, is perfectly rough inside and outside, and has inside it a rough solid cylinder of radius c. When the two are in motion on a perfectly rough horizontal table, prove that

$$\{(M+m)(3\,a^2+b^2)-ma^2\}\phi=m(b-c)\{b\theta+2\,a\sin\theta\}+Ct+C',$$

where M and m are the masses of the hollow and the solid cylinder respectively, ϕ the angle the hollow cylinder has turned through, and θ the angle which the plane containing their axes makes with the vertical after the time t.

282. A string of length c, fixed at one end, is tied to a uniform lamina at a point distant b from the centre of inertia. The centre of inertia is initially at the greatest possible distance from the fixed point and has a velocity v given to it in the plane of the lamina and perpendicular to the string. Prove that when the angle between the string and line b is a maximum, the angular velocity of the lamina is $\dfrac{v}{b+c}$ and the tension of the string is $\dfrac{2\,mv^2c}{(b+c)^2}\cdot\dfrac{K^2-2\,c(b+c)}{K^2-4\,c(b+c)}$, MK^2 being the greatest moment of inertia of the lamina at the centre of inertia.

283. In a circular lamina which rests on a smooth horizontal table and which can turn freely about its centre, which is fixed, a circular groove is cut. If a heavy particle be projected along the groove, supposed rough, with given velocity, find the time in which the particle will make a complete revolution (i) in space, (ii) relatively in the groove.

284. Four equal rods, each of mass m and length l, are connected by smooth joints at their extremities so as to form a rhombus. A constant force mf is applied to each rod at its middle point, and perpendicular to its length, — each force tending outwards. If the equilibrium of the system be slightly disturbed by pressing two opposite corners towards each other, and the system be then abandoned to the action of the forces, show that the time of a small oscillation in the *form* of the system is $2\pi\sqrt{\left(\dfrac{l}{3f}\right)}$.

285. A spherical shell of radius a and mass m rolls along a rough horizontal plane, whilst a smooth particle of mass P oscillates within the shell in the vertical plane in which the centre of the shell moves, the particle never being very far from the lowest point. Show that the time of its oscillation will be the same as that of a simple pendulum of length $= ma(a^2 + k^2) \div \{(m + P)a^2 + mk^2\}$, where k is the radius of gyration of the shell about a diameter.

286. A solid cylinder with projecting screw-thread is freely movable about its axis fixed vertically, and a hollow cylinder with a corresponding groove works freely about it without friction. Find the moment of the couple which must act on the solid cylinder in a plane perpendicular to its axis in order that the hollow cylinder may have no vertical motion.

287. A sphere rolls from rest down a given length l of a rough inclined plane, and then traverses a smooth part of the plane of length ml. Find the impulse which the sphere sustains when perfect rolling again commences, and show that the subsequent velocity is less than it would have been if the whole plane had been rough. In the particular case when $m = 120$, show that the velocity is less than it would otherwise have been in the ratio of 67 to 77.

288. A rough sphere is placed upon a rough horizontal plane which revolves uniformly about a vertical axis; the centre of

the sphere is attracted to a point in the axis of rotation, and in the same horizontal plane with itself by a force varying as the distance. Determine the motion.

289. A heavy uniform beam AB is capable of rotating in a vertical plane about a fixed axis passing through its middle point C, and is inclined to the vertical at an angle of 60°. If a perfectly elastic ball fall upon it from a given height, find how long a time will elapse before the ball strikes the beam again.

290. A sphere rests on a rough horizontal plane, half its weight being supported by an elastic string attached to the highest point of the sphere; the natural length of the string is equal to the radius and the stretched length to the diameter of the sphere. If the sphere be slightly displaced parallel to a vertical plane, show that the time of an oscillation is $\pi \sqrt{\left(\dfrac{14a}{15g}\right)}$.

291. A uniform heavy rod, movable about its middle point A, has its extremities connected with a point B by elastic strings, the natural length of each of which is equal to the length AB. Find the period of its small oscillations.

292. A squirrel is in a cylindrical cage and oscillating with it about its axis, which is horizontal. At the instant when he is at the highest point of the oscillation, he leaps to the opposite extremity of the diameter and arrives there at the same instant as the point which he left. Determine his leap completely.

293. A perfectly rough sphere rolls on the internal surface of a fixed cone, whose axis is vertical and vertex downwards. Prove that the angular velocity about its vertical diameter is always the same and that the projection on a horizontal plane of the radius vector of its centre, measured from the axis, sweeps out areas proportional to the times. Show also that the polar equation to the projection on a horizontal plane of the path of the centre is

$$7\frac{d^2u}{d\theta^2} + (2 + 5\sin^2\alpha)u = \frac{5g\sin\alpha\cos\alpha}{h^2u^2} - \frac{2c\gamma\cos\alpha}{h},$$

where α is the semi-vertical angle, γ the constant angular velocity about the vertical diameter, and h is half the area swept out by the radius vector in a unit of time.

294. A thin circular disc is set rotating on a smooth horizontal table, about a vertical axis through its centre perpendicular to its plane, with angular velocity ω in a wind blowing with uniform horizontal velocity v. Supposing the frictional resistance on a small surface a at rest to be $cvma$, where m is the mass of a unit of area, show that the angle turned through in any time is $\frac{\omega}{c}(1-e^{-ct})$, and that the centre of gravity moves through a space $vt-\frac{v}{c}(1-e^{-ct})$. Determine the same quantities for a frictional resistance $=cv^2ma$.

295. A uniform rod of length $2a$ passes through a small fixed ring, its upper end being constrained to move in a horizontal straight groove. Show that if the rod be slightly displaced from the position of equilibrium, the length of the isochronous simple pendulum will be $\dfrac{a^2+(a-b)^2}{3a}$, where b is the distance of the ring from the groove.

296. A homogeneous solid of revolution spins with great rapidity about its axis of figure, which is constrained to move in the meridian. Prove that the axis will oscillate isochronously, and determine its positions of stable and unstable equilibrium.

297. A wire in the form of the portion of the curve

$$r=a(1+\cos\theta),$$

cut off by the initial line, rotates about the origin with angular velocity ω. Show that the tendency to break at a point $\theta=\frac{\pi}{2}$ is measured by $\frac{1}{5}^2\sqrt{2}\cdot ma^3\omega^2$, where m is the mass of a unit of length.

298. Show that in every centrobaric body the central ellipsoid of inertia is a sphere. Is the converse of this proposition true?

299. A uniform sphere is placed in contact with the exterior surface of a perfectly rough cone. Its centre is acted on by a force the direction of which always meets the axis of the cone at right angles and the intensity of which varies inversely as the cube of the distance from that axis. Prove that if the sphere be properly started the path described by its centre will meet every generating line of the cone on which it lies at the same angle.

300. A sphere of radius a is suspended from a fixed point by a string of length l and is made to rotate about a vertical axis with an angular velocity ω. Prove that if the string make small oscillations about its mean position, the motion of the centre of gravity will be represented by a series of terms of the form $L \cos (kt + M)$, where the several values of k are the roots of the equation $(lk^2 - g)\left(k^2 - \omega k - \dfrac{5g}{2a}\right) = \dfrac{5gk^2}{2}$.

301. A rigid body is attached to a fixed point by a weightless string of length l, which is connected with the body by a socket (permitting the body to rotate freely without twisting the string) at a point on its surface where an axis through its centre of inertia, about which the radius of gyration is a maximum or a minimum, $= k$, meets it. The body is set rotating with angular velocity ω about such axis placed vertically (the string, which is tight, making an angle α with the vertical), and being then let go, show that it will ultimately revolve with uniform angular velocity

$$= \sqrt{\left\{\omega^2 + \frac{2g(1 - \cos \alpha)}{k^2}\right\}}.$$

302. Three equal uniform rods placed in a straight line are jointed to one another by hinges, and move with a velocity v perpendicular to their lengths. If the middle point of the middle rod become suddenly fixed, show that the extremities of the other two will meet in time $\dfrac{4\pi a}{9v}$, a being the length of each rod.

303. A top in the form of a surface of revolution, with a circular plane end, is set spinning on a smooth horizontal plane about its axis of figure, which is inclined at an angle u to the vertical. It is required to determine the motion and to show that the axis will begin to fall or to rise according as $\tan u >$ or $< \dfrac{b}{a}$, where b is the radius of the circular plane end perpendicular to the axis, and a is the distance of the centre of inertia from this end.

304. A heavy uniform beam AB of length a is capable of freely turning about the point A, which is fixed; the end B is suspended from a fixed point C by a fine inextensible chain of length c. The system being at rest is slightly disturbed. Find the time of a small oscillation, the weight of the chain being neglected.

Examine the case in which the line AC is vertical.

305. A perfectly rough sphere of radius a moves on the concave surface of a vertical cylinder of radius $a + b$, and the centre of the sphere initially has a velocity v in a horizontal direction. Show that the depth of its centre below the initial position after a time t is $\dfrac{5g}{2v^2}b^2(1 - \cos nt)$, where $n^2 = \dfrac{2v^2}{7b^2}$.

Show also that in order that perfect rolling may be maintained the coefficient of friction must not be less than $\dfrac{12\,bg}{7v^2}$.

306. A heavy particle slides down the tube of an Archimedian screw, which is vertical and capable of turning about its axis. Determine the motion.

WORKS ON PHYSICS, ETC.

PUBLISHED BY

THE MACMILLAN COMPANY.

AIRY. Works by Sir G. B. AIRY, K.C.B., formerly Astronomer-Royal.
On Sound and Atmospheric Vibrations. With the Mathematical Elements of Music. 12mo. $2.50.
Gravitation. An Elementary Explanation of the Principal Perturbations in the Solar System. 12mo. $1.90.

ALDIS: Geometrical Optics. An Elementary Treatise. By W. STEADMAN ALDIS, M.A. Third Edition, Revised. 12mo. $1.00.

DANIELL: A Text-Book of the Principles of Physics. By ALFRED DANIELL, D.Sc. Illustrated. 8vo. New revised edition, $4.00.

DAUBENY'S Introduction to the Atomic Theory. 16mo. $1.50.

DONKIN (W. F.): Acoustics. Second Edition. 12mo. $1.90.

EVERETT: Units and Physical Constants. By J. D. EVERETT, F.R.S. 16mo. $1.25.

FERRERS: Spherical Harmonics and Subjects Connected with them. By Rev. N. M. FERRERS, D.D., F.R.S. 12mo. $1.90.

FISHER: Physics of the Earth's Crust. By OSMOND FISHER. 8vo. $3.50.

FOURIER: The Analytical Theory of Heat. By JOSEPH FOURIER. Translated with Notes, by A. FREEMAN, M.A. 8vo. $4.50.

GALLATLY: Examples in Elementary Physics. Comprising Statics, Dynamics, Hydrostatics, Heat, Light, Chemistry, and Electricity. With Examination Papers. By W. GALLATLY, M.A. $1.00.

GARNETT: An Elementary Treatise on Heat. By W. GARNETT, M.A., D.C.L. Fifth Edition, Revised and Enlarged. $1.10.

GLAZEBROOK: Heat and Light. By R. T. GLAZEBROOK, M.A., F.R.S. *Cambridge Natural Science Manuals.* Cr. 8vo. $1.40. Bound separately. Heat, $1.00. Light, $1.00.

HEATH: Treatise on Geometrical Optics. By R. S. HEATH. 8vo. $3.50.
An Elementary Treatise on Geometrical Optics. 12mo. $1.25.

HOGG'S (JABEZ) Elements of Experimental and Natural Philosophy. $1.50.

IBBETSON: The Mathematical Theory of Perfectly Elastic Solids. By W. J. IBBETSON. 8vo. $5.00.

JELLETT (JOHN H. B. D.): A Treatise on the Theory of Friction. 8vo. $2.25.

JONES: Examples in Physics. By D. E. JONES, B.Sc. 16mo. 90 cents.
Sound, Light, and Heat. 16mo. 70 cents.
Lessons in Heat and Light. 16mo. $1.00.

LOEWY (B.): Experimental Physics. 16mo. 50 cents.
A Graduated Course of Natural Science, Experimental and Theoretical, for Schools and Colleges. 16mo. PART I. 60 cents. PART II. 60 cents.

LOUDON and McLENNAN: A Laboratory Course of Experimental Physics. By W. J. LOUDON, B.A., and J. C. McLENNAN, B.A., of the University of Toronto. 8vo. $1.90.

LOVE: Treatise on the Mathematical Theory of Elasticity. 8vo. Vol. I. $3.00. Vol. II. $3.00.

LUPTON: Numerical Tables and Constants in Elementary Science. By SYDNEY LUPTON. 16mo. 70 cents.

MACFARLANE: Physical Arithmetic. By ALEXANDER MACFARLANE, Professor of Physics, University of Texas. 12mo. $1.90.

MAXWELL: The Scientific Papers of James Clerk Maxwell, M.A., LL.D., B.Sc., etc., etc. Edited by W. D. NIVEN, M.A., F.R.S. With Steel Portraits and Page Plates. 2 vols. 4to. $25.00.

McAULAY (A.): Utility of Quaternions in Physics. 8vo. $1.60.

MOLLOY: Gleanings in Science. Popular Lectures on Scientific Subjects. By the Rev. GERARD MOLLOY, D.D., D.Sc. 8vo. $2.25.

NEWTON'S Principia. Edited by Professor Sir W. THOMSON and Professor BLACKBURN. (Latin Text.) 4to. $12.00.
    First Book of Newton's Principia. Sections I., II., III. With Notes and Problems. By P. FROST, M.A. Third Edition. 8vo. $3.00.
    The First Three Sections of Newton's Principia, with an Appendix; and the Ninth and Eleventh Sections. Fifth Edition edited by P. T. MAIN. $1.00.

NICHOLS: Laboratory Manual of Physics and Applied Electricity. Arranged and edited by EDWARD L. NICHOLS, Cornell University.
    VOL. I. Junior Course in General Physics. 8vo. $3.00.
    VOL. II. Senior Courses and Outlines of Advanced Work. 8vo. $3.25.

ORFORD: Lens Work for Amateurs. By HENRY ORFORD. 16mo. 80 cents.

PARKINSON: A Treatise on Optics. By S. PARKINSON, D.D., F.R.S. Fourth Edition, Revised and Enlarged. 12mo. $2.50.

PEARSON: A History of the Theory of Elasticity. By ISAAC TODHUNTER. Edited by Professor KARL PEARSON. Vol. I. Galilei to Saint-Venant, 1639–1850. 8vo. $6.00. Vol. II. Saint-Venant to Lord Kelvin (Sir William Thomson). In Two Parts. $7.50.

PERRY: An Elementary Treatise on Steam. By JOHN PERRY. With Woodcuts, Numerical Examples, and Exercises. 18mo. $1.10.

PRESTON (T.): The Theory of Light. 8vo. $5.00.
    The Theory of Heat. 8vo. $5.50.

RAYLEIGH: The Theory of Sound. By JOHN WILLIAM STRUTT, Baron RAYLEIGH, Sc.D., F.R.S. Second Edition, Revised and Enlarged. In two volumes. 8vo. Vol. I. $4.00.

SAINT-VENANT: The Elastic Researches of. Edited for the Syndics of the Cambridge University Press by KARL PEARSON, M.A. 8vo. $2.75.

SHANN: An Elementary Treatise on Heat in Relation to Steam and the Steam-Engine. By G. SHANN, M.A. 12mo. $1.10.

SHAW: Practical Work at the Cavendish Laboratory. Edited by W. N. SHAW. Heat. 8vo. 90 cents.

SPOTTISWOODE: Polarization of Light. By W. SPOTTISWOODE, LL.D. Illustrated. 12mo. $1.25.

STEWART. Works by BALFOUR STEWART, F.R.S.
    Lessons in Elementary Physics. 16mo. $1.10.
    Questions on the Same for Schools. By T. H. CORE. 40 cents.
    A Treatise on Heat. Sixth Edition. 16mo. $2.25.

STEWART and GEE: Lessons on Elementary Practical Physics. By BALFOUR STEWART, M.A., LL.D., F.R.S., and W. W. HALDANE GEE.
    VOL. I. General Physical Processes. 12mo. $1.50.
    VOL. II. Electricity and Magnetism. $2.25.
    VOL. III. Optics, Heat, and Sound. (*In the press.*)
    Practical Physics for Schools and the Junior Students of Colleges.
    VOL. I. Electricity and Magnetism. 16mo. 60 cents.
    VOL. II. Optics, Heat, and Sound. (*In the press.*)

STOKES (G. G.): **On Light.** On the Nature of Light. On Light as a Means of Investigation. On the Beneficial Effects of Light. 12mo. $2.00.
Mathematical and Physical Papers. 8vo.
VOL. I. $3.75. VOL. II. $3.75. VOL. III. (*In the press.*)

STONE: **Elementary Lessons on Sound.** By W. H. STONE, M.B. With Illustrations. 16mo. 90 cents.

TAIT. Works by P. G. TAIT, M.A., SEC. R.S.E.
Lectures on Some Recent Advances in Physical Science. With Illustrations. Third Edition, Revised and Enlarged, with the Lecture on Force delivered before the British Association. 12mo. $2.50.
Heat. With Numerous Illustrations. 12mo. $2.00.
Light. An Elementary Treatise. With Illustrations. 12mo. $2.00.
Properties of Matter. Second Edition, Enlarged. 12mo. $2.25.

TAYLOR: Sound and Music. By SEDLEY TAYLOR, M.A. Illustrated. Second Edition. 12mo. $2.50.
Theories of Sound. Paper. 10 cents.

THOMSON. Works of J. J. THOMSON, Professor of Experimental Physics in the University of Cambridge.
A Treatise on the Motion of Vortex Rings. 8vo. $1.75.
Application of Dynamics to Physics and Chemistry. 12mo. $1.90.

THOMSON. Works of Sir W. THOMSON, F.R.S., Professor of Natural Philosophy in the University of Glasgow.
Mathematical and Physical Papers.
VOL. I. 8vo. $5.00. VOL. II. 8vo. $4.50. VOL. III. 8vo. $5.50.
Popular Lectures and Addresses.
VOL. I. **Constitution of Matter.** 12mo. $2.00.
VOL. II. **Geology and General Physics.** 12mo. $2.00.
VOL. III. **Navigational Affairs.** With Illustrations. 12mo. $2.00.
On Elasticity. 4to. $1.25.
On Heat. 4to. $1.25.

TODHUNTER: A History of the Theory of Elasticity. By ISAAC TODHUNTER. Edited by Professor KARL PEARSON.
VOL. I. Galilei to Saint-Venant, 1639–1850. 8vo. $6.00.
VOL. II. Saint-Venant to Lord Kelvin (Sir William Thomson). In Two Parts. $7.50.

TURNER: A Collection of Examples on Heat and Electricity. By H. H. TURNER, B.A. 12mo. 75 cents.

WALKER: The Theory and Use of a Physical Balance. By JAMES WALKER, M.A. 8vo. 90 cents.

WATSON and **BURBURY: A Treatise on the Application of Generalized Co-ordinates to the Kinetics of a Material System.** By H. W. WATSON, D.SC., and S. H. BURBURY, M.A. 8vo. $1.50.

WOOD: Light. By Sir H. TRUMAN WOOD. 16mo. 60 cents.

WOOLCOMBE: Practical Work in Heat. By W. G. WOOLCOMBE, M.A., B.SC. Crown 8vo. $1.00.
Practical Work in General Physics. Crown 8vo. 75 cents.

WRIGHT: Light. A Course of Experimental Optics, Chiefly with the Lantern. By LEWIS WRIGHT. 12mo. New Edition, Revised. $2.00.

ELECTRICITY AND MAGNETISM.

ALLSOP (F. C.): **Practical Electric Light Fitting.** $1.50.
BENNETT: The Telephoning of Great Cities. Paper. 35 cents.
BLAKESLEY: Alternating Currents of Electricity. Third Edition. Enlarged. (*In the press.*)

BONNEY (G. E.): **Induction Coils.** $1.00.
 Electrical Experiments. 12mo. 75 cents.
BOTTONE (S. R.): **Electricity and Magnetism.** 16mo. 90 cents.
 How to Manage the Dynamo. 16mo. 60 cents.
 A Guide to Electric Lighting. 16mo. 75 cents.
CAVENDISH: **Electrical Researches of the Honourable Henry Cavendish, F.R.S.** Written between 1771 and 1781. $5.00.
CHRYSTAL: **Electricity, Electrometer, Magnetism, and Electrolysis.** By G. CHRYSTAL and W. N. SHAW. 4to. $1.60.
CUMMING: **An Introduction to the Theory of Electricity.** By LINNÆUS CUMMING, M.A. With Illustrations. 12mo. $2.25.
DAY: **Electric Light Arithmetic.** By R. E. DAY, M.A. 18mo. 40 cents.
EMTAGE: **An Introduction to the Mathematical Theory of Electricity and Magnetism.** By W. T. A. EMTAGE, M.A. 12mo. $1.90.
GRAY: **The Theory and Practice of Absolute Measurements in Electricity and Magnetism.** By ANDREW GRAY, M.A., F.R.S.E. In two volumes. Vol. I., 12mo. $3.25. Vol. II. (in two parts). $6.25.
 Absolute Measurements in Electricity and Magnetism for Beginners. By ANDREW GRAY, M.A., F.R.S.E. Students' Edition. 16mo. $1.25.
GUILLEMIN: **Electricity and Magnetism.** By AMÉDÉE GUILLEMIN. Translated and edited, with Additions and Notes, by Professor SILVANUS P. THOMPSON. Super royal 8vo. $8.00.
HAWKINS and **WALLIS**: **The Dynamo.** Its Theory, Design, and Manufacture. By C. C. HAWKINS and F. WALLIS. $3.00.
HEAVISIDE (OLIVER): **Electrical Papers.** 2 vols. 8vo. New Edition. $7.00.
HERTZ (H.): **Researches in the Propagation of Electrical Forces.** Authorized Translation by D. E. JONES, B.Sc. Illustrated. 8vo. $3.00.
JACKSON (D. C.): **A Text-Book on Electro-Magnetism and the Construction of Dynamos.** By DUGALD C. JACKSON, B.S., C.E., Professor of Electrical Engineering, University of Wisconsin. 12mo. $2.25.
 Alternating Currents and Alternating Current Machinery. (*In the press.*)
LODGE (OLIVER J.): **Modern Views of Electricity.** By OLIVER J. LODGE, LL.D., D.Sc., F.R.S. Illustrated. $2.00.
 Lightning Conductors and Lightning Guards. With numerous Illustrations. $4.00.
MAXWELL: **An Elementary Treatise on Electricity.** By JAMES CLERK MAXWELL, M.A. Edited by WILLIAM GARNETT, M.A. 8vo. $1.90.
 A Treatise on Electricity and Magnetism. By JAMES CLERK MAXWELL, M.A. 2 vols. 8vo. Second Edition. $8.00.
 Supplementary volume, by J. J. THOMSON, M.A. $4.50.
MAYCOCK: **Electric Lighting and Power Distribution.** By W. P. MAYCOCK. In one volume, $1.75. In three parts, 16mo, paper, 75 cents each.
 A First Book of Electricity and Magnetism. Crown 8vo. 60 cents.
POOLE: **The Practical Telephone Handbook.** By JOSEPH POOLE. $1.00.
PREECE and **STUBBS**: **A Manual of Telephony.** By WILLIAM HENRY PREECE and ARTHUR J. STUBBS. $4.50.
PRICE: **A Treatise on the Measurement of Electrical Resistance.** By WILLIAM A. PRICE. 8vo. $3.50.
RUSSELL: **Electric Light Cables and the Distribution of Electricity.** By STUART A. RUSSELL, A.M., I.C.E. 12mo. $2.25.
STEWART and **GEE**: **Practical Physics for Schools and Junior Students of Colleges.** By BALFOUR STEWART, M.A., LL.D., F.R.S., and W. W. HALDANE GEE, B.Sc.
 Vol. I. Electricity and Magnetism. 16mo. 60 cents.

Lessons in Elementary Practical Physics. By BALFOUR STEWART, M.A., LL.D., F.R.S., and W. W. HALDANE GEE, B.SC.
Vol. II. Electricity and Magnetism. 12mo. $2.25.

THOMSON: **Notes on Recent Researches in Electricity and Magnetism.** Intended as a Sequel to Professor CLERK MAXWELL'S "Treatise on Electricity and Magnetism." By J. J. THOMSON. 8vo. $4.50.
Elements of the Mathematical Theory of Electricity and Magnetism. 12mo. $2.60.

THOMSON: **Reprints of Papers of Electrostatics and Magnetism.** By Sir WILLIAM THOMSON, D.C.L., LL.D., F.R.S., F.R.S.E. 8vo. $5.00.

THOMPSON: **Elementary Lessons in Electricity and Magnetism.** By SILVANUS P. THOMPSON, D.SC., B.A., F.R.A.S. New Edition. With Illustrations. 16mo. $1.40.

WALKER: **How to Light a Colliery by Electricity.** 4to. Limp. 75 cents.
Town Lighting by Electricity. (*In the press.*)

WATSON and **BURBURY**: **The Mathematical Theory of Electricity and Magnetism.** By H. W. WATSON, D.SC., F.R.S., and S. H. BURBURY, M.A.
VOL. I. Electrostatics. 8vo. $2.75.
VOL. II. Magnetism and Electrodynamics. 8vo. $2.60.

WHETHAM. **Solution and Electrolysis.** By W. C. D. WHETHAM, Trinity College, Cambridge. Cr. 8vo. $1.90.

MECHANICS.

ALDIS: **Rigid Dynamics, An Introductory Treatise on.** By W. STEADMAN ALDIS, M.A. $1.00.

ALEXANDER: **Model Engine Construction.** By J. ALEXANDER. 12mo. Cloth. pp. xii, 324. Price, $3.00.

ALEXANDER and **THOMPSON**: **Elementary Applied Mechanics.** PART II. Transverse Stress. $2.75.

BALL: **Experimental Mechanics.** A Course of Lectures delivered to the Royal College of Science for Ireland. By Sir R. S. BALL, LL.D., F.R.S. Second Edition. With Illustrations. 12mo. $1.50.

BASSET: **A Treatise on Hydrodynamics.** 2 vols. 8vo. $9.00.
An Elementary Treatise on Hydrodynamics and Sound. 8vo. $3.00.
A Treatise on Physical Optics. 8vo. $6.00.

BAYNES: **Lessons on Thermodynamics.** By R. E. BAYNES, M.A. 12mo. $1.90.

BESANT: **A Treatise on Hydromechanics.** Fifth Edition, Revised. PART I. Hydrostatics. 12mo. $1.25.
A Treatise on Dynamics. $1.75.
Elementary Hydrostatics. 16mo. $1.00. KEY. (*In the press.*)

CLIFFORD: Works by W. KINGDON CLIFFORD, F.R.S.
Elements of Dynamic. An Introduction to the Study of Motion and Rest in Solid and Fluid Bodies.
PART I. Books I.–III. $1.90. PART II. Book IV. and Appendix. $1.75.

COTTERILL: **Applied Mechanics.** An Elementary General Introduction to the Theory of Structures and Machines. By J. H. COTTERILL, F.R.S. $5.00.

COTTERILL and **SLADE**: **Elementary Manual of Applied Mechanics.** By Prof. J. H. COTTERILL, F.R.S., and J. H. SLADE. 12mo. $1.25.

CREMONA (LUIGI): **Graphical Statics.** Two Treatises on the Graphical Calculus and Reciprocal Figures in Graphical Calculus. Authorized English Translation by T. HUDSON BEARE. 8vo. $2.25.

GARNETT: **Elementary Dynamics, A Treatise on.** For the Use of Colleges and Schools. By WILLIAM GARNETT, M.A., D.C.L. Fifth Edition. $1.50.

GLAZEBROOK: **Mechanics.** By R. T. GLAZEBROOK, M.A., F.R.S. Crown 8vo. In parts: Dynamics, $1.25. Statics, 90 cents. Hydrostatics. (*In the press.*)

GOODWIN: **Elementary Statics.** By H. GOODWIN, D.D. 75 cents.

GREAVES: Works by JOHN GREAVES, M.A.
A Treatise on Elementary Statics. 12mo. $1.90.
Statics for Beginners. 16mo. 90 cents.
Treatise on Elementary Hydrostatics. 12mo. $1.10.

GREENHILL: Hydrostatics. By A. G. GREENHILL. 12mo. $1.90.

GUILLEMIN (A.): The Applications of Physical Forces. Translated and Edited by J. NORMAN LOCKYER, F.R.S. With Colored Plates and Illustrations. Royal 8vo. $6.50.

HICKS: Elementary Dynamics of Particles and Solids. By W. M. HICKS. 12mo. $1.60.

HOROBIN: Elementary Mechanics. By J. C. HOROBIN, B.A. With Numerous Illustrations. 12mo. Cloth. Stages I. and II., 50 cents each. Stage III. 50 cents.
Theoretical Mechanics. Division I. (*In the press.*)

HOSKINS: The Elements of Graphic Statics. A Text-book for Students of Engineering. By L. M. HOSKINS, C.E., M.S. 8vo. $2.25.

JELLETT: A Treatise on the Theory of Friction. By JOHN H. JELLETT, B.D., late Provost of Trinity College, Dublin. 8vo. $2.25.

JESSOP: The Elements of Applied Mathematics, including Kinetics, Statics, and Hydrostatics. By C. M. JESSOP. $1.25.

KENNEDY: The Mechanics of Machinery. By ALEXANDER B. W. KENNEDY, F.R.S. With Illustrations. 12mo. $3.50.

LAMB: Hydrodynamics. A Treatise on the Mathematical Theory of Fluid Motion. By H. LAMB. 8vo. New Edition enlarged. $6.25.

LOCK: Works by the REV. J. B. LOCK, M.A.
Dynamics for Beginners. 16mo. $1.00.
Elementary Statics. 16mo. $1.10. KEY. 12mo. $2.25.
Mechanics for Beginners. PART I. 90 cents. Mechanics of Solids.
Elementary Hydrostatics. (*In preparation.*)
Mechanics of Solids. 16mo. (*In the press.*)
Mechanics of Fluids. 16mo. (*In the press.*)

LONEY: A Treatise on Elementary Dynamics. New and Enlarged Edition. By S. L. LONEY, M.A. 12mo. $1.90.
SOLUTIONS OF THE EXAMPLES CONTAINED IN THE ABOVE. 12mo. $1.90.
The Elements of Statics and Dynamics.
PART I. Elements of Statics. $1.25.
PART II. Elements of Dynamics. $1.00.
Complete in one volume. 12mo. $1.90. KEY. 12mo. $1.90.
Mechanics and Hydrostatics for Beginners. 16mo. $1.25.

MACGREGOR: An Elementary Treatise on Kinematics and Dynamics. By JAMES GORDON MACGREGOR, M.A., D.Sc. 12mo. $2.60.

MINCHIN: Works by G. M. MINCHIN, M.A.
A Treatise on Statics. Third Edition, Corrected and Enlarged.
VOL. I. Equilibrium of Coplanar Forces. 8vo. $2.25.
VOL. II. Statics. 8vo. $4.00.
Uniplanar Kinematics of Solids and Fluids. 12mo. $1.90.
Hydrostatics and Elementary Hydrokinetics. $2.60.

PARKER: A Treatise on Thermodynamics. By T. PARKER, M.A., Fellow of St. John's College, Cambridge. $2.25.

PARKINSON (R. M.): Structural Mechanics. $1.10.

PARKINSON: A Treatise on Elementary Mechanics. For the Use of the Junior Classes at the University and the Higher Classes in Schools. With a collection of Examples by S. PARKINSON, F.R.S. 12mo. $2.25.

PIRIE: Lessons on Rigid Dynamics. By the Rev. G. PIRIE, M.A. 12mo. $1.50.

RAWLINSON: Elementary Statics. By G. RAWLINSON, M.A. Edited by E. STURGES. 8vo. $1.10.

ROUTH: Works by E. J. ROUTH, LL.D., F.R.S.
A Treatise on the Dynamics of a System of Rigid Bodies. With Examples. New Edition, Revised and Enlarged. 8vo. In Two Parts.
PART I. Elementary. Fifth Edition, Revised and Enlarged. $3.75.
PART II. Advanced. $3.75.
Stability of a Given State of Motion, Particularly Steady Motion. 8vo. $2.25.
A Treatise on Analytical Statics. With Numerous Examples. Vol. I. 8vo. $3.75.

SANDERSON: Hydrostatics for Beginners. By F. W. SANDERSON, M.A. 16mo. $1.10.

SELBY: Elementary Mechanics of Solids and Fluids. $1.90.

SYLLABUS OF ELEMENTARY DYNAMICS.
PART. I. Linear Dynamics. With an Appendix on the Meanings of the Symbols in Physical Equations. Prepared by the Association for the Improvement of Geometrical Teaching. 4to. 30 cents.

TAIT and **STEELE:** A Treatise on Dynamics of a Particle. By Professor TAIT, M.A., and W. J. STEELE. Sixth Edition, Revised. 12mo. $3.00.

TAYLOR: Resistance of Ships, and Screw Propulsion. By D. W. TAYLOR. $3.75.

TODHUNTER. Works by ISAAC TODHUNTER, F.R.S.
Mechanics for Beginners. With Numerous Examples. New Edition. 18mo. $1.10. KEY. $1.75.
A Treatise on Analytical Statics. Fifth Edition. Edited by Professor J. D. EVERETT, F.R.S. 12mo. $2.60.

WALTON: Mechanics, A Collection of Problems in Elementary. By W. WALTON, M.A. Second Edition. $1.50.
Problems in Theoretical Mechanics. Third Edition Revised. With the addition of many fresh Problems. By W. WALTON, M.A. 8vo. $4.00.

WEISBACH and **HERRMANN:** The Mechanics of Hoisting Machinery, including Accumulators, Excavators, and Pile-Drivers. A Text-book for Technical Schools, and a Guide for Practical Engineers. By Dr. JULIUS WEISBACH and Professor GUSTAV HERRMANN. Authorized Translation from the Second German Edition. By KARL P. DAHLSTROM, M.E., Instructor of Mechanical Engineering in the Lehigh University. With 177 Illustrations. $3.75.
Mechanics of Pumping Machinery. (*In the press.*)

ZIWET: An Elementary Treatise on Theoretical Mechanics. In Three Parts: Kinematics, Statics, and Dynamics. By ALEXANDER ZIWET, University of Michigan. 8vo. Complete in one volume. $5.00.
PART I. $2.25.   PART II. $2.25.   PART III. $2.25.

THE MACMILLAN COMPANY,
NEW YORK: 66 FIFTH AVENUE.
CHICAGO: ROOM 23, AUDITORIUM.

A LABORATORY MANUAL

OF

EXPERIMENTAL PHYSICS.

BY

W. J. LOUDON and J. C. McLENNAN,

Demonstrators in Physics, University of Toronto.

Cloth. 8vo. pp. 302. $1.90 *net.*

FROM THE AUTHORS' PREFACE.

At the present day, when students are required to gain knowledge of natural phenomena by performing experiments for themselves in laboratories, every teacher finds that as his classes increase in number, some difficulty is experienced in providing, during a limited time, ample instruction in the matter of details and methods.

During the past few years we ourselves have had such difficulties with large classes; and that is our reason for the appearance of the present work, which is the natural outcome of our experience. We know that it will be of service to our own students, and hope that it will be appreciated by those engaged in teaching Experimental Physics elsewhere.

The book contains a series of elementary experiments specially adapted for students who have had but little acquaintance with higher mathematical methods: these are arranged, as far as possible, in order of difficulty. There is also an advanced course of experimental work in Acoustics, Heat, and Electricity and Magnetism, which is intended for those who have taken the elementary course.

The experiments in Acoustics are simple, and of such a nature that the most of them can be performed by beginners in the study of Physics; those in Heat, although not requiring more than an ordinary acquaintance with Arithmetic, are more tedious and apt to test the patience of the experimenter; while the course in Electricity and Magnetism has been arranged to illustrate the fundamental laws of the mathematical theory, and involves a good working knowledge of the Calculus.

THE MACMILLAN COMPANY,
66 FIFTH AVENUE, NEW YORK.